AIDS in Africa

AIDS in Africa:
Its Present and Future Impact

Tony Barnett and Piers Blaikie

The Guilford Press
New York London

Published by the Guilford Press
A Division of Guilford Publications, Inc.
72 Spring Street, New York, NY 10012

Printed in Great Britain

This book is printed on acid-free paper

Last digit is print number: 9 8 7 6 5 4 3 2 1

Library of Congress Cataloging-in-Publication Data

A CIP record for this book is available from the Library of Congress

ISBN 0-89862-880-6
Library of Congress Catalog Card Number 91- 077918

Typeset by BookEns Ltd. Baldock, Herts.
Printed and bound in Great Britain

Contents

List of Figures

List of Tables

Preface

This has been a difficult book to research and write. Studying the impact of a disease such as AIDS on other people was always bound to be distressing. It was only when we confronted the details of disrupted lives in Uganda, and later when we reviewed field notes and other data, that the full weight of the situation became apparent. Thus our deepest thanks and acknowledgements must go to the ordinary people of Rakai and Kigesi Districts in Uganda who welcomed us into their communities and talked with us when they had many more practical and pressing problems on their minds. In addition, the District Administrator of Rakai District, Mrs E. R. Kassadha, was of the greatest help.

We have incurred many professional debts and we wish to express our thanks to the following individuals: Dr Dan Mudoola of Makerere Institute of Social Research, who gave us wise advice and extended much in the way of practical assistance in establishing a research base in Uganda; Dr Susan Hunter of Makerere University, who shared her insights and her research data with enormous and unusual generosity (unfortunately we have been unable to make full use of the final report of this research (Dunn et al., 1991) as it only came into our possession as this book was going to press), and her husband Mr Arlin Green, whose good humour and friendship contributed to our visits to Uganda; Dr Christine Obbo, the project anthropologist, who worked for many months in the trying circumstances of rural Rakai, who collected some of the project data (but who must bear no responsibility for its interpretation), and whose linguistic skills were important for the translation of some of the interview and written material; Dr Sam Kalibala of TASO, who took time out from his medical work in Masaka in order to stay in some of the villages and to extend his own knowledge while sharing it with us; Ms Abby Nalwanga-Sebina, Dr Sam Zziwa and Mr Dyshan Musakweta, who worked with us as research associates in the field for seven weeks and contributed greatly to this project through their hard work and constant good humour; Dr Stan Musgrave of Columbia University, who discussed his own work in Rakai with us; and Dr Maryinez Lyons who in discussing her own research plans with us directed us to much useful literature.

The research would never have been carried out had it not been for the imagination and foresight of the Economic and Social Committee for Overseas Research of the Overseas Development Administration of the British Government in deciding to fund a programme of research on the impact of AIDS - a decision which, in 1988, was very foresightful indeed. Needless to say, neither the Overseas Development Administration nor the Government of Uganda can be held responsible for any of the data or interpretation contained in this book, although both were fully apprised of the findings in 1990. Within the ODA, special thanks are due to Dr Rosalind Eyben who was instrumental in commissioning the research, and to Mr Peter Skinner who administered it, as well as to Dr David Nabarro who took a keen interest in what we were doing. At the British High Commission in Kampala, the then First Secretary, Mr Mike Hammond, was unusually helpful and encouraging, as was Mrs Trish Cavell who arranged logistic support, in particular transport.

At the University of East Anglia, the School of Development Studies and the Overseas Development Group provided the usual supportive if sometimes - at a time of cuts in university funding throughout the United Kingdom - stressful working environment. The interdisciplinary nature of the School always enabled us to call upon colleagues for assistance when we were uncertain of our bearings. Special thanks are due to Ms Jane Bartlett of the Overseas Development Group who administered the project at the University of East Anglia, and to Mr John Maxwell who drew the maps and diagrams.

There are, of course, more personal debts. Piers Blaikie wishes to thank Sally Blaikie for her tolerance of both his physical and mental absence from her life, while Tony Barnett is deeply indebted to his wife, Sarah Knights, who has endured both of these kinds of absence as well as his constant preoccupation with this work over the last three years.

And, finally, great thanks to Mr Michael Johnson who came and repaired the computer when it suffered catastrophic memory failure at the very last moments of the preparation of this manuscript!

Tony Barnett
Piers Blaikie

School of Development Studies
University of East Anglia
Norwich, United Kingdom
September 1991

1

The Downstream Impact of AIDS in Africa

Epidemic diseases may have profound social effects. AIDS (Acquired Immune Deficiency Syndrome) is such a disease. Its impacts will differ from one society to another. In both poor and rich countries it will strain health-care budgets; in some poor countries, such as those of Africa, it may reduce the numbers of adults who can produce food and care for the young and the old.

Fear of some social groups as especially prone to a disease results in stigma, never more so than when the disease is sexually transmitted and fatal. AIDS is mainly a sexually transmitted disease (STD). Its transmission is via body fluids that harbour the Human Immunodeficiency Virus (HIV).

In the early 1980s, doctors in the United States noticed increasing frequency of an unusual form of pneumonia. A medical report called simply and obscurely 'Pneumocystis pneumonia – Los Angeles' heralded recognition of this life-threatening epidemic in June 1981 (*Morbidity and Mortality Weekly Report*, 5 June 1981, pp. 250–52). The sudden appearance of many cases of this and also of a rare cancer – ***Kaposi's sarcoma*** – indicated that something unusual was afoot. Patients were dying because their immune systems were unable to combat these and other, more common, illnesses. The first victims were young homosexual men. Records of prescriptions for the drug pentamidine used to treat the pneumonia showed few cases prior to 1979, suggesting an increasing and unusual prevalence of the condition. If there was an epidemic, what was it an epidemic of? Early suggestions for a name indicated the prejudices so easily associated with the disease. GRIDS – Gay Related Immune Deficiency Syndrome – was a first attempt. Cases were soon found in people other than homosexual men. In 1982 the United States Centers for Disease Control coined the name AIDS. In Africa, the magnitude of the epidemic became apparent in the latter half of 1986, the number of cases reported to the World Health Organization rising from 31 in May 1986 to 2,627 by February 1987 (Piot and Mann, 1987; Fleming, 1988a). Identification of the disease agent and its mode of transmission presented a challenge to medical science in the 1980s. In particular, a test that could identify people infected with HIV, or a drug to delay its onset or which would provide a cure, offered fame and fortune to medical researchers. Identification of the HIV gave rise to intense competition between research institutes, culminating in a dispute between teams from the Institut Pasteur in France and the National Cancer Institute in the USA, a dispute which gave rise to a long-running and only recently resolved feud (Connor and Kingman, 1988; Gallo, 1991; Specter, 1991).

HIV is a slow-acting virus able to reproduce itself using genetic material from the cells of its host. As with many other viral disease agents, such as the common cold, it readily mutates, making the development of a vaccine or a treatment very difficult indeed. It can lie dormant for many years, enabling infectious but asymptomatic people to appear healthy. This aspect of the disease means that many people may be infected before medical, social and political responses can be mobilized. It is present in every country, some of the highest rates being reported from the USA, France, Brazil and, in Africa, the Republic of Congo, Rwanda and Uganda. In Africa, the disease has been spread heterosexually from the outset. The rapidity of its spread can be partly explained by the lack of health resources, poor general health, long periods of social unrest and economic disruption. Levels of infection are so high in some countries – in Uganda about eight per cent of the population may be infected – that the number of deaths over the next decades will almost inevitably slow rates of population increase and in some particularly heavily infected areas, it will result in population decline (Rowley, Anderson and Wan Ng, 1990; Anderson, May, Boity, Garnett and Rowley, 1991). This has serious social and economic implications. A disease mainly affecting people aged between 15 and 50 years of age will result in large numbers of orphans, shortages of labour, the loss of expensively trained specialists, as well as increasing the burden of health and other forms of care.

Such effects are not restricted to the countries of Africa. In the rich world, too, hard decisions will have to be made about the allocation of resources between the care of people with AIDS and the care of people with other illnesses. In parts of South East Asia, inequalities between rich and poor countries, between men and women and between urban and rural areas, produces 'sex tourism' and a fertile environment for the spread of the disease. Thus James Chin, of the World Health Organization's Global Programme on AIDS predicted at the Seventh International AIDS Conference in Florence, June 1991, that in the next five years the number of infected people in Asia could rise to three million – most of them in India and Thailand. As Asia has almost a third of the world's population, the number of people at risk is very large indeed.

Condom use is one suggested way of slowing the epidemic and some political and religious groups will have to accept both their use and a franker discussion of sexual matters. However, it has to be recognized that recommendations to use condoms may fail because of the practical difficulties of obtaining and using them in communities where there is little money or privacy, much sexual modesty and perhaps, above all, because this is a method over which women inevitably have little control and which men dislike.

Certain things are clear about AIDS. It is a disease that can affect anybody. It is not especially a disease of gays, black people or of intravenous drug users. It is not even exclusively sexually transmitted. In Romania it is widespread among young children who have received blood transfusions. Its origins are obscure, and although it is widely believed by some that the most commonly occurring form of the virus, HIV-1, originated in Africa, there is currently no clear evidence to support this hypothesis (Fleming, 1990, p. 177), although recent research points increasingly to that possibility (Anderson, May, Boily, Garnett, Rowley, 1991, p. 586). Suggestions that it is 'caused' by any particular group invite moral panic rather than the careful judgements and effective policies that will save lives. Knowledge of the origin and spread of HIV is important for the understanding of its genetic make-up and might contribute to a strategy for dealing with it. However, in a world where prejudice and discrimination along ethnic and sexual lines is widespread, such knowledge if misinterpreted threatens another epidemic of prejudice and discrimination (Chirimuuta and Chirimuuta, 1987; Hooper, 1990, pp. 219–23).

The Human Immunodeficiency Virus destroys the body's defence mechanisms. It does not kill people directly; it opens the way for other infections that do kill as the body becomes decreasingly able to muster its defences. Thus people die not directly from the HIV but from the effects of other, sometimes normally mild infections which abound in the environment and to which their compromised immune systems allow them to fall prey. The virus is fragile. It cannot live for very long outside the human body and passes from person to person via the medium of body fluids such as blood, semen and vaginal secretions. Thus it is not exclusively a sexually transmitted disease, but may usually be. This means that in addition to being deadly it also carries with it many of the stigmatizing and morally resonant associations of a sexually transmitted disease.

Disease does not only affect the physical body. It also affects the 'social body', the relationships between people. As with any other illness, AIDS makes people dependent, less able to play their part in their family or household. It may put them into a condition of socially defined 'impurity' (Douglas, 1966). This is never more so than when a disease resonates with people's sexuality. Indeed, in the nineteenth century, such diseases were known as 'social diseases'. A disease such as AIDS which is understood socially as being both sexual and life-threatening is likely to be socially disruptive in the extreme. Two observers (Connor and Kingman, 1988) have described it as a disease of the devil and as we shall see, in Uganda the frightening and socially disruptive aspects of the disease are captured by some of the names by which people describe it – 'the robber', 'the one that drains', 'the cheater', 'the incurable disease which imprisons us'. Because of its particularly threatening nature, it is also a disease which rapidly becomes socially defined as a disease of 'the impure other' – affecting some culturally defined out-group, homosexuals, black people, foreigners, prostitutes. While at first the epidemic may affect some population sub-groups more than others, the passing of time will ensure that its distribution within the population alters. Thus what was, in North America and Europe, once seen as a disease of male homosexuals, haemophiliacs and intravenous drug users, is increasingly becoming a disease to which every sexually active person may be exposed, or which may affect those undergoing or practising surgery or dentistry.

Globally, the majority of people now infected by the HIV is heterosexual, and globally the majority of those who will develop AIDS in the future will also be heterosexual. In June 1991, James Chin said that by the year 2000 around 90 per cent of the global AIDS cases would be found in the general heterosexual population. He continued: 'The point is to say that the future of the HIV and AIDS pandemic is in the heterosexual population' (The *Guardian*, 18 June 1991, p. 5).

This realization is important as a background to the present book; so is the realization that this is not a disease that is restricted to or can be identified with a particular sexual orientation or ethnic group. The way in which some epidemiologists describe and classify the particular pattern of a stigmatizing disease such as AIDS may itself reflect the same social and cultural prejudices that have made the disease shameful in the first place. In a world where for historical and political reasons, difference can be, and often has been, interpreted as evaluation of inferiority and superiority between supposed 'racial' or ethnic groups, it is important to be particularly careful not to reinforce social prejudice in the name of science. This has already occurred with AIDS – labelled as the 'gay plague' or identified as 'an African disease'. It has been cogently argued that AIDS policy is best understood using the models developed in response to crime and disorder as opposed to sickness and health (Small, 1988). The idea of the dangerousness of those who are associated with high levels of infection is central here, and one which elicits moral panic. The association of male homosexuals, drug users and black Africans and Caribbeans as high-risk groups in the white Western press – and it seems in parts of the academic and international research community, too, as exemplified below – can serve to push the problem into a ghetto.

An example will serve to illustrate this point. At the first International Conference on the Global Impact of AIDS, held in London in 1988, two similar papers were presented. Each paper examined the distribution of AIDS cases within ethnically heterogeneous national populations. The two national populations were Brazil and the USA (Rodrigues, 1988; Selik et al., 1988). These papers were interesting as much for the different ways in which they approached the same problem of describing a population as for the substantive material they contained. The American paper described the distribution of the illness in terms of ethnic group and sexual orientation, while the Brazilian paper did the same task in terms of socio-economic class and sexual orientation. The American paper thus gave the impression that AIDS was predominantly a disease of 'blacks' and

'Hispanics', while the Brazilian paper suggested that it was mainly a disease of poor people whose poverty drove them to earn their living in ways that exposed them to infection. In each of these cases the key descriptive classification chosen by the researchers reflected the 'commonsense' view of what were considered relevant ways of distinguishing between population sub-groups in their own societies. This example of the way that the work of scientists may be formed and influenced by their own social beliefs is of particular importance in a book that describes the impact of a predominantly sexually transmitted disease in Africa.

What is of the greatest importance in this book is not that it describes the impact of a predominantly sexually transmitted disease on Africans. Rather, it describes the impact of a fatal disease that happens to be sexually transmitted on people who happen to be African. The real lesson of the book is that this is a disease that is likely to have major social and economic effects on any society. In Africa, these effects reflect the circumstances of culture, income and ways of making a living, which are common on that continent. In Europe or North America different ways of life, cultural orientations and levels of life will mean that the impacts are specific to these parts of the world, while in those parts of South East Asia that provide sexual tourism for visitors from Europe and North America, the form of the epidemic will be different again. But, as will be seen in what follows, the human experience of illness and death is universal.

The solution to the AIDS problem will come from medical science, but the effects of the disease will be felt by individuals, communities and societies for years to come. In the absence of a cure or a vaccine, millions of individuals will die. The importance of writing about Africa is that on that continent the disease is spreading very rapidly in impoverished communities which depend on human labour for survival and where the levels of national poverty are already so great that the resources for dealing with the care of the sick and dying and the orphans are already extremely scarce. The total health budget of Zaïre, with a population of about 28 million people is, for example, less than the annual budget for a district general hospital in the USA or the UK. In such countries in particular the loss of labour resulting from the epidemic is likely to have substantial 'downstream' impacts on the society and economy as subsistence farming communities, commercial plantations, mines, factories and the administration lose people to the disease.

Against this background, in Africa the AIDS pandemic confronts us with the full range of development issues. Thus we shall see that issues of poverty, entitlement and access to food, medical care and income, the relationships between men and women, the relative abilities of states to provide security and services for their peoples, the relations between the rich and the poor within society and between rich and poor societies, the viability of different forms of rural production, the survival strategies of different types of household and community, all impinge upon a consideration of the ways in which an epidemic such as this affects societies and economies.

But this is not all. This disease may affect every society in the world.

There are few countries in which cases of AIDS have not been reported. Thus we see that the implications of the disease are international and transcultural. To reiterate the point, this is not a disease of gay people, black people, white people or of people who use drugs. It is simply a disease of people. Many of the ways in which the disease is transmitted, contracted and coped with by patients, their relatives and society as a whole are transcultural and will be familiar to any one of us, regardless of our origins. They include many relations of intimacy and care - sexual relations, pregnancy and birth, blood transfusion and medical treatment - as well as the darker side of human experience - drug abuse, prostitution and rape.

Studying the social and economic impact of AIDS

This book is about the social and economic consequences of AIDS in Africa, with special reference to Uganda, where the authors were involved in a research project for about 18 months. Its writing was preceded by two years of search and research for funds and information. The fact that these activities were fraught with difficulties is only to be expected, but the nature of the difficulties were themselves instructive because they derived at least in part from the nature of the research topic.

The AIDS pandemic shares some of the characteristics of other pandemics such as bubonic plague, measles, smallpox and Asian influenza. These other pandemics developed in ways that were the outcome of modes of transmission and pathology particular to the diseases themselves and of the ways in which people lived their daily lives. AIDS, as an event that impacts in a decisive and adverse way upon a population, can be studied by social scientists in terms of the concept of a *disaster*. This is the approach adopted in this book.

Understanding a disaster involves two sets of considerations. The first is physical specifications of the hazard itself and the second the characteristics of the population at risk. When the hazard is a disease, some of the relevant aspects are its symptoms, mode of transmission, infectiousness, incubation period and the range of outcomes in those infected. When the hazard is a drought, the amount of available moisture and its timing is, among other climatic and agronomic data, one of the main considerations. With regard to the population at risk, a wide range of social data become relevant when confronted with the particular characteristics of the hazard. In the case of drought, the farming system, the extent of local food stocks and the abilities to command entitlements to food, including cash reserves and non-agricultural employment, are examples of relevant social data. Behind these proximate measures of vulnerability to the effects of drought lie more pervasive and deep-seated socioeconomic structures. Divisions along the lines of class, age, gender, caste and tribe will set the context of particular vulnerabilities to the particular hazard in question.

In the same way, AIDS can be studied exclusively as a medical phenomenon. Such an approach emphasizes the physical characteristics of the hazard. In the same way, too, the risk of contracting the disease by the population can be explained by the degree of exposure to the main modes of infection through the interchange of bodily fluids via sexual intercourse, receiving contaminated blood and so on. These are the proximate or immediate causes of infection and many of them relate to 'sexual practice'. However, behind these practices lie the ways in which a particular society structures sexual relations - the expectations of marriage, cohabitations and other liaisons, and of favours received and given in sexual encounters whether hetero- or homosexual. Here the analysis moves away from medical science to epidemiology, and, in the study of these contextual structures, to sociology and anthropology. In this book we have adopted the paradigm of contemporary disaster theory. We hope that by doing so we are able to make clear the possible downstream effects of such a disaster over the longer term.

So, if the authors set out to study the impact of AIDS in terms of other recognized disasters and diseases, why did they face so many difficulties in the initial stages? The first reason was that AIDS was, and is, an *unprecedented* disaster - the modes for studying it had not been thought through and some of its implications had not occurred to potential research funding bodies. Just as Uganda and other societies are learning by experience what to expect and how to adapt existing modes of coping with the risk and the downstream impacts of the disease, so planners, international agencies and holders of research funds likewise had to have time to adapt established ways of thinking towards more appropriate action. The formation of the Global Programme on AIDS (GPA) within the World Health Organization forced that institution to think afresh about many of its entrenched views on preventive medicine, intravenous drug use and so on (Connor and Kingman, 1988). In turn, the GPA, under the leadership of Jonathan Mann, began to sensitize national governments to the urgent necessity of recognizing the problem and of acting quickly to mitigate its effects on health. Similarly, the United Nations' Food and Agriculture Organization (FAO) first saw the AIDS pandemic as a medical problem, and not as one which might affect food production in the longer term. These perspectives over-emphasized the physical aspect of the hazard and neglected the socially and economically determined issues of vulnerability and the effects of such a pandemic on economy and society - particularly the longer term downstream effects such as discussed in this book. However, times have changed. The FAO has now initiated a research programme to study the effects of AIDS upon food production, and some of their results are reviewed in what follows.

This need to adapt to the unprecedented nature of the AIDS pandemic brings us to the next difficulty of researching the impact of AIDS. It has primarily been researched as a medical problem. After all, it is medical science which will eventually provide a vaccine or successful treatment of the disease. The seemingly endless series of international conferences in

Washington, London, Montreal, Stockholm, San Francisco, Naples, Marseilles, Kinshasa and most recently at Florence, have been dominated by medical scientists. However, because AIDS is a disaster that like any other requires full consideration of the vulnerability of the population at risk, its study also requires the *sharing* of research concerns between medical and social scientists. If this is not done, then only one arm of the pincer movement, that of the physical nature of the hazard, or the socioeconomic characteristics of the population at risk, receives attention. This partial and inadequate treatment of a disaster will be familiar to those working in other areas of disaster research, characterized by a preoccupation with building design and seismology in the case of earthquakes, and meteorology in the case of drought. These considerations are, it hardly needs to be said, crucial elements in understanding a disaster, but they are incomplete if they do not take into account the socioeconomic determinants and outcomes of vulnerability to the physical hazard.

In the case of AIDS research, the cooperation and shared programme of research into what best should be done has extended partly into the social sciences, but for a limited and specialized purpose. This concerns the epidemiological aspects of sexual practice. Predicting the spread of the disease and identifying the most sensitive parameters within sexual practice has stimulated considerable research effort. For Africa the work of Caldwell et al. (1989), of Larson (1989) and the extensive review of sexual behaviour by Standing and Kisekka (1989), is instructive. However, even in this dialogue between epidemiology and sociology and anthropology, there seem to be difficulties of communication between separate academic cultures. The authors witnessed an example of this at a conference in the UK on the role of anthropology in the study of AIDS where a plea for 'anthropological data' to provide inputs to mathematical epidemiological model-building was met by antipathy on the part of some anthropologists to what they considered useless number crunching, when the 'real' problem was one of understanding the world views of people with AIDS and people living in AIDS-affected communities. This antipathy and lack of understanding of the need of constructive multi-disciplinary dialogue is, we have found, not only in that direction. At the present time, the number of social science studies of AIDS is still quite small. Biritwum et al. (1989) found that of 551 AIDS research projects considered, only 86 were sociobehavioral, and most of these concerned the identification of risk groups by means of the most proximate variables (for example, the presence of co-factors in infection rates, the number of sexual partners and knowledge-attitude-practice). Hardly any researchers had studied the socioeconomic structures determining either 'upstream' vulnerability – the risk of being infected – or 'downstream' vulnerability – the risks associated with poor or no treatment and the impact upon the household in its economically productive as well as its social aspects.

Since the start of our research, institutions and individuals have made adaptations in order to meet the unprecedented challenges that AIDS presents to professional, research and political cadres throughout the world. In many ways, all of us have started to adapt in a similar fashion to those at risk in Uganda – the people who are the main focus of this book – trying known solutions to unfamiliar problems and experimenting with new ideas. New research initiatives have begun under the auspices of the Social Science Research Council of the United States, USAID, the Ford Foundation, the Food and Agriculture Organization of the United Nations, the Overseas Development Administration of the United Kingdom, as well as at least a dozen research and academic establishments in the Anglophone world. As this book will show, there are sociologists and economists contributing to multidisciplinary projects on AIDS. There is a growing literature on the wider determinants of vulnerability to infection and the impact of AIDS. As the pandemic develops, the downstream effects, often unforeseen a few years ago, are receiving more attention. Many politicians have abandoned the usual impulse to deny or shift the blame on to others when unpleasant events appear on their horizon, and have instead started to develop policies and structures to combat the problem. Many of these learning processes have now been embarked upon by researchers as well as by people and communities affected by the disease.

AIDS as a development issue

The World Health Organization recognizes three general patterns of AIDS (Mann, 1988) and a summary account and world map (Figure 1.1) is virtually mandatory introductory material for books on this subject. Three patterns are recognized. In pattern I countries, HIV began to spread extensively in the 1970s. Most cases were homosexual or bisexual males and urban intravenous drug users. Heterosexual transmission is now increasing, and in some countries new heterosexual HIV cases exceed all others. The male to female ratio is in the range 10:1 to 15:1. Pattern II countries are characterized by predominantly heterosexual transmission and thus the male to female ratio is about 1:1, although there is increasing African evidence that more women are infected than men (Chin, 1990; Berkley et al., 1990; Anderson, May, Boily, Garnett, Rowley, 1991, p. 582). Drug user and homosexual or bisexual transmission is very rare, but contaminated blood and blood products are still a problem in countries that have not yet implemented nation-wide screening. Extensive spread started about the same time as in Pattern I countries. Finally, Pattern III countries experienced AIDS and HIV cases much later and it is likely that the virus was introduced in the mid 1980s. Many fewer cases have been reported, although there is now evidence of very rapid increase in Thailand and India (W. Carswell, personal communication).

All three patterns can be found on the African continent (Anonymous (1), 1987; Fleming, 1988a, 1988b; Fleming, 1990). Pattern I occurs mainly among white male homosexuals in South Africa. Pattern II can be seen in two large areas of Africa: in a belt from Gabon in the west across to the

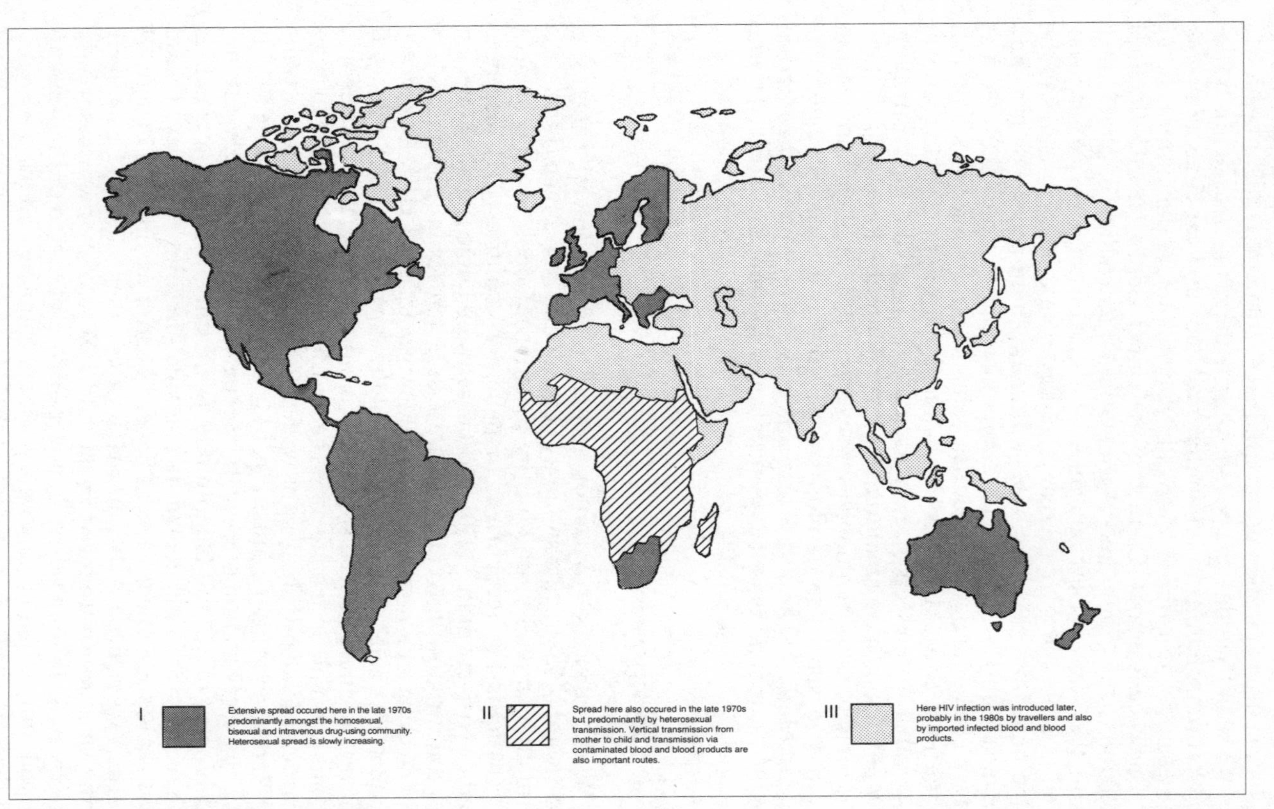

Figure 1.1 Global patterns of AIDS

Sources: Wellcome, 1989; Chin, Sato and Mann, 1989.

east coast, and from Uganda to Zimbabwe, with concentrations around the great lakes and highlands of central Africa. The epidemic of HIV-II (a different and possibly less virulent strain of the virus) is to be found in West Africa from the Cape Verde Islands to Benin, with high incidences in Southern Senegal and Guinea Bissau on the Atlantic seaboard. Africa north of the Sahara currently exhibits pattern III, but it is suspected that significant transmission of HIV-1 may have started in many of these countries too (Benslimane et al., 1987).

However, in general, Africa, as can be seen from Figure 1.1 is a Pattern II area. As Chapter 2 will discuss in more detail, it is estimated that sub-Saharan Africa has 64 per cent of the global total of AIDS cases, and projections of future numbers in Africa are for greater increases than for any other region in the world. Although the USA has 51 per cent of reported cases, and the highest case rate per head of population according to official statistics, African populations are the most vulnerable to future infection, inadequate treatment of sufferers, and social and economic disruptions that might give the epidemic added impetus. Effective policy measures in the fields of public education, the basic medical care of sufferers and the treatment of other diseases that increase the rate of transmission ('co-factors' such as genital ulcer disease), the legal infrastructure to protect sufferers in their workplace and the dependants of mortality cases, and social security for sufferers and their dependants - all these have been developed only partially in many African countries. For this reason AIDS becomes a development issue, an issue with implications for almost every aspect of that process.

Take a typical case. A woman stands in front of her house in Uganda. Her husband died of AIDS three months ago. Aged 45, Nankya now lives alone with her children. She is, by local standards, fairly well off - in a country where per capita Gross National Product was estimated at $220 in 1985 (Kaijuka et al, 1989, p. 2). But all is not well. Her husband's relatives have taken her to court to try and gain control of the land they insist is rightfully theirs. For the moment she has an half hectare plot of green bananas (matooke), the staple food in Buganda, as well as others of beans, sweet potatoes, cassava, tomatoes, chili peppers and Irish potatoes. This is a fertile region of Africa, blessed with good soils and abundant rainfall. Even in the most adverse political conditions that have characterized the last 20 years, food production and supply has rarely been a problem. But how long can this be the case in the face of a disease that threatens to destroy the labour supply upon which the survival of subsistence households depends? Can she support her children? Will she be able to keep the land? Above all, will she too soon die of AIDS?

Nankya is in need of many things: medical care, money to educate her children, labour to help her on her farm, legal protection to maintain her tenure of her husband's land, knowledge that she will receive a proper burial, assurance as to the welfare of her children after her death. As a citizen of Uganda, living in a rural area, she can be sure of few of these. For example, central government expenditure on health in Uganda had fallen from 1.16 per cent of GDP in 1972 to just 0.23 per cent by 1986, which represents $0.64 per head, or about $2.00 if funding of the AIDS Control Programme, sponsored by WHO, is included (World Bank, 1987). As a member of a community in which there are many people affected by the disease, she knows that other households are under the same types of stress; they may not be able to extend the accepted types of solidarity and material assistance that have been available in times of individual stress or disaster.

In the early 1970s, this part of Uganda enjoyed what was arguably the best rural infrastructure in Africa. For example, in the early 1970s only a mile from Nankya's home there was a rural clinic, complete with solar-powered refrigerator and well-stocked dispensary, with an adjacent maternity hospital. Today, after decades of civil insecurity and warfare, a period during which the very structure of the state has at times almost ceased to exist, the buildings remain but they are in disrepair. The refrigerator is no longer operational and there are virtually no drugs. So, Nankya does not have access to medical care.

Primary and secondary schools in the village near her home have survived 20 years of civil and economic disorder. School children can be seen being taught under the trees, the buildings are still standing, most of the teachers remain. However, there is no money for maintenance, the teachers themselves often go without pay, which is meagre when they receive it. Responsibility for resourcing this rural school has increasingly moved away from government and has fallen on the shoulders of the parents and the local community. So Nankya will have to find 3,600 shillings[1] per child at the beginning of each term in addition to the compulsory school uniform and books, pencils, rulers and all the other little things that enable a child to benefit from primary and secondary education to the full and not appear disadvantaged to their peers.

As we have seen, Nankya grows a wide range of food crops on her farm. Now that her husband is dead, and she is becoming less capable of working long hours herself, she must substitute fewer labour-intensive crops since she cannot put in the time necessary to grow a range of crops and provide a varied diet for her household. Sometimes she has to find workers to help her – all this after a year of diverting her labour time from the farm to the care of her dying husband. However, although she is not among the poorest in this community she does not have sufficient cash (for this can only be obtained from paid work or from the sale of crops or handicrafts) to pay her labourers. Her teenage son who is well capable of helping her does not do so. He says that this kind of work is not for a man, and in any case he plans to move to Kampala and become wealthy. For men and women, the work they do, the resources to which they have access and the rewards, psychological as well as material, they can expect, lie at the heart of any society. In this society, women are expected to shoulder the major burden of care as well as of subsistence production.

Nankya's hold on her major productive asset – her land – is subject to disputes arising from gender biases in the system of land tenure in present day Uganda. The legal position concerning land tenure is fraught with confusion and uncertainty because of overlapping and mutually inconsistent land laws and customs. Such uncertainties can be exploited to the disadvantage of women and children in particular, those who are least able to protect themselves through the courts. In Nankya's case, as we have seen, her deceased husband's brother now threatens to dispossess her.

In all societies and for most ordinary people, children and funerals are the only expectation they may have of surviving in people's memories after the grave. In Buganda, funeral ceremonies are elaborate. It is here that the value of individuals and their position in and contribution to their lineage, their clan and their neighbours is attested by the community. Such occasions are both costly and time-consuming. Neither money nor time is freely available. The AIDS pandemic has killed so many in this community that people now frequently have to choose between time spent farming or time spent attending up to four funerals a week. As far as money is concerned, not only is this a poor community, but the epidemic itself has had a disastrous effect on some households. In Nankya's case, she has already paid for the funerals of her husband who died from AIDS, as well as of her infants who died from other causes. In Uganda, for every 1,000 live births, 101 children die in their first year and another 88 do not live to the age of five (Kaijuka et al, 1989, p. 54).

Finally, Nankya knows that she will not long survive her husband. So the future of the children, who will be orphaned, becomes uncertain. Orphaning is not uncommon in this society. Life expectancy in Uganda is 46.9 years for women and 45.6 years for men (Kaijuka et al, 1989, p. 2). Orphaning is usually dealt with effectively by orphans going to join the household of a close relative. Indeed, in the past Nankya's children have spent long periods in their dead father's brother's household. But he is already looking after the orphaned children of another brother and is unable to take on any further responsibility. The progress of the epidemic in this area has meant that this effective and flexible system of coping is showing signs of breakdown. Nankya's own parents are too old. Her sisters live far away and she has lost touch with them. She can only hope that her neighbours and friends will provide on-site assistance to the children in their own home. Since there is no system of social security provided by the government, she must hope that some kind of satisfactory arrangement can be made.

Even before the process of coping with the impact of the disease has begun, the social and spatial distribution of the disease through time also confronts important development issues. Each society shapes its own particular AIDS epidemic. This is not the truism it may appear. AIDS is a disease transmitted by means of body fluids. In sub-Saharan Africa it is predominantly transmitted through heterosexual intercourse. So the incidence of HIV and AIDS reflects the social, economic and cultural circumstances of sexual relations. Other epidemics of fatal disease have affected particular classes or other social categories more than others. Few other fatal epidemic diseases have been so age- and behaviour-specific in their pattern of transmission. AIDS is most likely to affect people between 15 and 45 years of age and can be described as highly age-cohort specific.

This book is about how people cope with AIDS in one country in Central Africa. Although its main substance concerns Uganda, there is, we believe, much here that is relevant to everyone - governments, policy-makers and the public as a whole throughout Africa and in other countries - as they confront a future with AIDS. By the year 2000, the total number of people in the world who are HIV+ may be as high as 20 million. The WHO estimates that as many as half a billion people are at risk worldwide (Brown, 1991, p. 17). So we have to recognize that this is a global problem, a human problem, but one which will have different effects on societies differing in their histories, cultures, levels and ways of life. To return to the view from the World Health Organization. James Chin has said that: 'Although all projection must be interpreted cautiously, there can be no doubt that during the next several decades, AIDS in most developing countries will become the leading cause of death among adults in their most productive years, and will also be one of the leading causes of infant and child mortality in many regions.' He also commented: 'in sub-Saharan Africa it was now estimated that one in 40 men and one in 40 women carried the virus' (The *Guardian*, 18 June 1991, p. 5.) Given these figures and what follows in this book, it will be apparent that the situation in some regions of Africa is already far worse. The case of Uganda today may present an image of the future both for other parts of the continent and elsewhere. The current director of the Global Programme on AIDS summed up these and other issues when he said in a recent interview: 'We have to think of AIDS as a development problem, not just a health one.' (The *Courier*, March–April 1991.)

Note

[1] In 1990, the official exchange rate was about 250 shillings to the pound sterling; the parallel (or black market) rate was nearer two thousand.

A note on further reading

There is a vast and increasing technical literature on all aspects of AIDS, although comparatively little of it is concerned with the social and economic aspects of its impact. Useful general reading includes: Connor and Kingman, 1988 (on the science of the disease), Hooper, 1990 (full of information about both the early impact of the pandemic in Uganda and the difficulties of researching it); while Shilts, 1987, is a massively detailed and highly readable account of the politics of AIDS in the USA.

2

The Demographic Impact of AIDS in Africa

AIDS in Africa is likely to result in exceptional and unusual demographic changes in some areas over the medium term and thus have serious socioeconomic effects. In poor subsistence-based societies where the disease transmission is predominantly heterosexual, and to a lesser extent from infected mothers to their children, the actual and potential socioeconomic effects require careful monitoring if some of the more severe downstream effects are to be mitigated or avoided. We do not claim that the processes of social and economic change described in this book are happening in all the countries of sub-Saharan Africa affected by the pandemic in the same way as is the case in Uganda. However, it is very likely that similar processes will be occurring in some parts of those countries.

Estimating the scale of the pandemic

This chapter discusses the possible present and future scale of the AIDS pandemic in Africa with a specific focus on Uganda. Since the focus of the book is upon the downstream impact of the disease, a detailed review of the large and growing literature on the demographic changes brought about by AIDS will not be attempted here. While it is argued that much of the quantitative information on present mortality and seroprevalence[1] is clouded with uncertainty, it is becoming clearer that the pandemic threatens major demographic changes for some regions. First, the sources of uncertainty are examined. This is followed by a review of the nature of the threat.

There are three interrelated problems in estimating present and future levels of seropositivity and mortality from AIDS in Africa. These are: inadequacies of data collection on mortality from AIDS and on seropositivity; irresponsible reportage and generalization from non-representative samples; and, largely but not wholly deriving from these problems, the absence of technical understanding and lack of experience gained through time or from analogous historical epidemics, of the AIDS pandemic itself.

The first issue concerns the collection of data on AIDS mortality. Although many African countries have established surveillance procedures for this purpose, and the situation has improved dramatically in recent years, there are understandably many reasons for under-reporting. Most surveillance systems involve passive reporting from existing hospitals and health posts. Reporting of AIDS cases usually underestimates actual numbers for several reasons. These include: mis-diagnosis, reporting fatigue by returning health units, incomplete or missing returns especially from outlying and under-staffed centres, and, most significant of all, people with AIDS not reporting to any returning health care facility. Now that the majority of African countries has a surveillance system of some sort in operation, the situation regarding AIDS cases has improved. For example, Uganda set up a surveillance system in 1987, which involved the development and testing of a clinical case definition for AIDS and a reporting form and training for District Medical Officers and their assistants (Berkley et al., 1988). Nonetheless, under-reporting is inevitable: estimates that this may be in the order of a factor of ten for Uganda cannot be substantiated and must remain hearsay, although rough calculations of real totals can be made from serosurveys and epidemiological modelling which lend support to this estimated level of under-reporting.

Statistics about seroprevalence are much more useful for estimating future morbidity and mortality. Large numbers of serosurveys have been carried out in different countries of Africa, but they have largely been based upon fairly small samples of non-representative groups such as high risk, urban-based prostitutes, women bar staff, truck drivers, blood donors and pregnant mothers. All except the last type of sample are unsuitable for extrapolating seroprevalence to the general population. National serosurveys are expensive to mount and require considerable expertise and laboratory facilities, yet it is upon such representative and competently executed surveys that reliable estimates must be based.

Information about levels of seroprevalence is sometimes difficult to obtain because of widespread unprofessional, unscientific and unco-operative practices among researchers from the medical and academic professions and officials from national governments and international agencies. This results, it seems, in part from reversion to racially-derived prejudice by expatriates (Chirimuuta and Chirimuuta, 1987). There also exist mutually suspicious relations between Western 'commando' scientists, who use African case material to further their careers rather than to assist in developing indigenous capabilities, and highly sensitive national officials, who feel themselves to be patronized and cynically manipulated (Hooper, 1990). There has also undoubtedly been a reluctance on the part of some national governments to admit the seriousness of the problem for economic reasons related to the potential damage to their tourist industries, and to a number of other political reasons. Fortunately, this aspect of denial at the national level has markedly diminished during the past few years. Ankrah (1989) and Serawadda and Katongole-Mbidde (1990) both write about the shortcomings of expatriate researchers in Uganda. From personal experience in researching AIDS in Africa and personal communications with many others engaged in the same activities, it seems that the level of mutual suspicion between workers of all origins is pervasive enough to inhibit on occasions the development of properly conducted research and the free sharing of data.

The modelling of the demographic impact of AIDS in Africa has therefore to cope with the sometimes questionable accuracy and limited quantity of available data. There are well-developed epidemiological modelling techniques to predict the future demographic impact of AIDS, but they demand an accuracy of data that simply does not exist in Africa – and to a lesser degree throughout the world. Chin and Mann (1989) have outlined the difficulties of long-term prediction. These include: the short period for which it has been possible to follow the pandemic (about eight years); a lack of analogous pandemics upon which to base estimates of parameters; a lack of knowledge about sexual practices and their association with transmission of the virus; few credible estimates of numbers of people with different sexual practices, levels of sexual activity, partner change, etc.; the rate of progression of the disease from initial infection to death; the relative role of co-factors (e.g. other sexually transmitted diseases) in facilitating infection; the changing degree of infectiousness of HIV-infected persons in the course of their illness; and the fact that most data refer to HIV-1 and not HIV-2, which has now established itself in West Africa. A more detailed discussion of the difficulties of model building can be followed in Anderson et al. (1988 and 1991). Bongaarts states categorically that 'a lack of information about the rate of spread of HIV-1 and its determinants in specific countries makes it impossible accurately to forecast the size of the epidemic in different countries' (Bongaarts, 1989, page reference not available – information retrieved from database).

Such a brief, underqualified review of the difficulties involved in obtaining accurate estimates of the present and future demographic impact of AIDS must not lead us to treat all data and subsequent analysis on AIDS in Africa with distrust. There is now a cumulative database on both mortality and seroprevalence in Africa as a result of the initiative of the Global Programme on AIDS, part of whose brief was to help national governments to establish reporting systems and national AIDS policies. This was a major step which forced policy makers and politicians to confront the issue and to evaluate the seriousness of the situation on the basis of scientifically, routinely collected information.

WHO also counselled the careful use of these statistics and that credence should not be lent to unsubstantiated and unrepresentative reports that would discredit all information collecting activities and create unfounded panic and inappropriate action (for example, fierce discrimination against and scapegoating of HIV-infected people). While this was undoubtedly the only way forward in the development of policies based on facts, it did lead to a clear divergence between official statistics on the one hand, which most observers suspected of gross under-reporting, particularly in the early days, and, on the other hand, 'corridor talk' in international conferences on AIDS and unofficially at meetings of NGO's and other points of contact in the capital cities of Africa, of very much higher levels of infection and mortality. Such information was, of course, unofficial, sometimes unreliable, but cumulatively credible.

However, it is perhaps true to say that this divergence has narrowed for a number of reasons. Firstly, national data collection systems have improved and more and better serosurveys have been carried out. Secondly, many nations have decided to 'go public': the most recent case is Zimbabwe, which released startling results of HIV tests on 1,000 pregnant women in Harare which suggest that 28.5 per cent of the active workforce may be HIV-positive (*Africa Analysis*, 17 May 1991, p. 14). Thirdly, both medical and socioeconomic research have enabled epidemiologists to refine their estimates of some of the parameters, which Chin and Mann (1989) had identified as particularly difficult to quantify.

The final and most unfortunate reason concerns the development of the pandemic itself. In certain areas of Africa and in certain occupational groups, it has made its presence suddenly felt. The epidemiological curve for the pandemic, while differing from others such as measles, shares the general form of a long tail, where there are few cases of HIV-positivity (which might well be missed in even quite large serosurveys), followed by a rapid increase in infected people (Anderson et al., 1988). This stage has been reached in many parts of eastern and central Africa and in certain occupational and income groups elsewhere. Such large numbers showing up in serosurveys and as ill patients in hospitals cannot be ignored; they confirm that the pandemic is taking a predicted pattern through space and time. Simply, it is easier to predict the course of a pandemic at a later stage than near its beginning when projections are based upon small numbers of infective people as well as on unreliable data.

AIDS related mortality and seroprevalence

Thus a detailed and reliable quantitative account of the present and future demographic impacts of AIDS in Africa is not possible at present. However, clearer patterns are cumulatively emerging, and more reliable glimpses of the present and future scale of the pandemic are now possible, so that orders of magnitude for future impacts can now be estimated over the short and medium term, and some credibility given to different scenarios over the next 20 years or so.

It is worth repeating that Africa faces a dearth of both contemporary and historical information on AIDS-related mortality and HIV seroprevalence. Also, any quantitative estimates of the present situation will tend to become quickly outdated. For example, the doubling time of seroprevalence levels has been estimated at between nine to eighteen months for some areas in central Africa. Thus, any data presented here are indicative of patterns and trends and should not be assumed to describe the situation at the present time.

Figure 2.1 indicates the reported and estimated share by continent of the global number of AIDS cases – those people who are exhibiting the final stages of the disease and not those who have actually died of it. It can be seen that Africa claims the greatest share of cases and that the under-reporting there gives a misleading impression of the seriousness of the disease in that continent. It is also estimated that about 10 per cent of infections have occurred from infected mothers to their foetus or infant; this source of infection will probably become the biggest cause of infant and child mortality in many developing countries.

There have been a number of attempts to model the future demographic impact of AIDS worldwide. WHO sponsored a projection, using the Delphi technique. This involved a process whereby WHO developed its own forecasting model, invited a number of experts to participate in using it, provided them with assumptions on which the model rested and issued a questionnaire to the participants as to their best guesstimates of the values to be given to the variables in the model. A series of negotiations took place to agree upon the different predictions provided by the participants and to provide a composite predictive model for AIDS cases until the year 2000. The result of this exercise is shown in Figure 2.2.

The main assumptions of these predictions are as follows:

1 the number of AIDS cases will increase by at least five and up to ten fold over the next five years;
2 there will be about three million AIDS cases worldwide by the year 2000 as a result of infection before the baseline date of 1988;
3 if there is assumed to be no preventative action through education programmes and other improvements in health care, there will be over six million AIDS cases by the year 2000;
4 over a million AIDS cases can be prevented worldwide by preventive measures such as public education, (although the calibration of this aspect of the model is little more than a guess).

The value of this projection three years after it was carried out is not so much for the actual numbers (we shall see in the case of Africa that the predictions that contributed to the global total are already bound to be far too conservative). Rather, it is useful as an indicator of future patterns. Greatly increased numbers of cases are inevitable and the rate of increase at this aggregated level shows no sign of declining. This conclusion is based on the assumption that practices that cause infection will not be amenable to marked change in the short term. Also, the form of the epidemiological curve follows a characteristic 'S' shape and the middle stages of the pandemic are characterized by rapid rates of growth in infection.

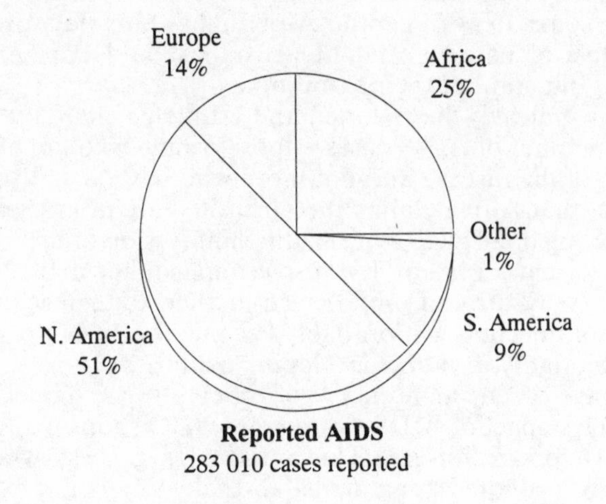

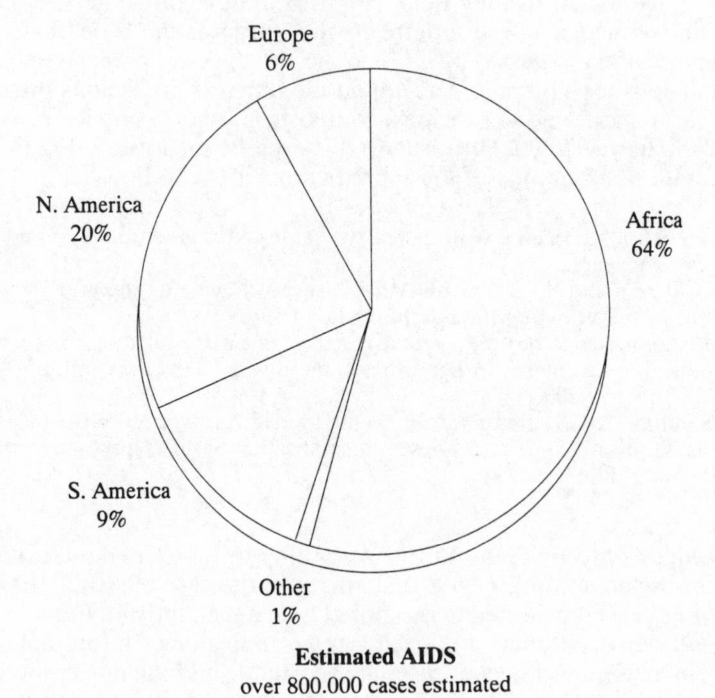

Figure 2.1 Reported and estimated AIDS cases, September 1990

Source: The Courier, 1991.

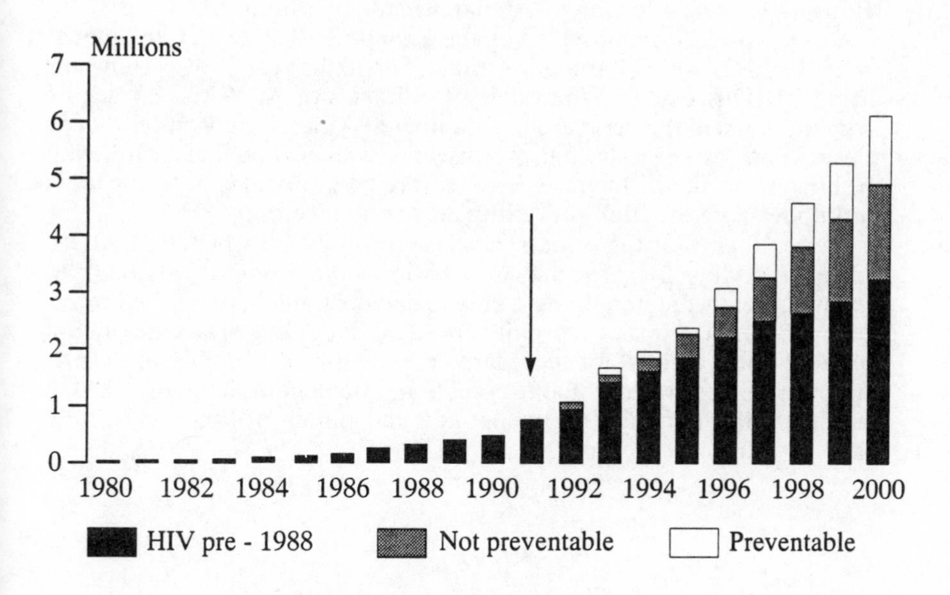

Figure 2.2 WHO Delphi projection of global AIDS

Source: Chin, Sato and Mann, 1989; The *Courier*, 1991.

Global estimates, while they have a mobilizing role in raising international consciousness, do not tell us anything about the greatly varying degree of seriousness of the pandemic from one region or category of people to the next. Instead, it is more useful to focus attention on different regions of the world. For example, as was mentioned in Chapter 1, Pattern I, II and III countries each have greatly differing rates of infection, modes of transmission and capabilities in preventing future infection. Africa, as discussed in Chapter 1, is a Pattern II region (heterosexual transmission is the chief mode of transmission). The Delphi Projection assumed a cumulative total of 150,000 cases at the time of projection (mid-1988), with an HIV prevalence of 2.5 million, of whom 500,000 can be expected to die in the next five years from the date of publication (Chin, Sato and Mann, 1989). In high-prevalence populations within Africa, the mortality rate in adults aged between 20 and 50 years could exceed the death rate from all other causes. There is also estimated to have been a cumulative total of 500,000 HIV-infected children born in Africa. Child-mortality rates in Africa are currently between 80-140 per thousand live births and this could be increased in the short term by between 10 per cent and 50 per cent. Some of these estimates are probably too conservative, since they themselves show large increases from estimates made only a year before (Bongaarts, 1988, referring to estimates made by Mann, 1988).

Another projection, provided by the Canada East West Centre (quoted in Lewis, 1991), which has made estimates up to the year 2000, is shown in Table 2.1. This explores the range of estimates of AIDS-related deaths from the worst to the best case in a number of African countries. Either of these scenarios suggests that the disease poses formidable health-care problems, but the differences between the two estimates are very large and underscore the high prevailing degree of uncertainty.

We see then that there are a number of projections of future AIDS-related mortality for Africa. It is not within the scope of this book to review critically their credibility, only to demonstrate that on most projections the implications of mortality are such that unless a vaccine or cure is found and provided for these large populations it is highly likely that the pandemic will bring about exceptional demographic change in the medium term, although the magnitude and timing of this may still be open to doubt.

Table 2.1 Estimates of AIDS-related deaths in the 15 to 50 age groups of selected Anglophone African countries

	1987 - 1991				1991 - 2000			
	Best Case (000)	**%**	**Worst Case (000)**	**%**	**Best Case (000)**	**%**	**Worst Case (000)**	**%**
Kenya	14	0.1	35	3.2	700	4.0	4,320*	26.9
Tanzania	52	0.4	1,300	11.0	1,180*	7.1	4,720*	28.4
Zimbabwe	3	0.1	65	1.4	130	2.0	1,040	15.8
Ghana	3	0.0	85	1.2	170	1.7	1,360	13.7
Zambia	12	0.3	300	8.3	360*	7.3	1,440*	29.4
Uganda	52	0.6	1,300	15.9	820*	7.2	3,280*	28.8
Total	136	0.3	3,085	6.7	3,360	5.1	16,160	24.8

Notes:

1 Data in the percentage columns represent AIDS-related deaths as a percentage of the 15 to 50 age group. The assessed age group accounts for about 45 per cent of the total population in most African countries.

*These numbers would be higher were it not for the assumption that no more than 50 per cent of the assessed age group will contract HIV.

Source: Lewis, 1991

Present differential mortality and HIV infection with special reference to Uganda

The distribution of mortality and HIV infection in Africa, as elsewhere, shows important elements of differentiation. There are four main parameters to be considered. The first is geographical, at the national, rural-urban, sub-national and micro-regional scales. The second is differential infection and mortality by occupational group and related to this, income and educational status. The third is gender and the fourth is age. These components do cross cut and they are not mutually exclusive. Therefore they are discussed one at a time, but it must be borne in mind that they overlap.

Starting with the variation in AIDS cases between the nations of Africa, Figure 2.3 shows the reported cases per million, using 1990 statistics. The case rates shown here even for the worst affected countries are very low and within error bands for most mortality statistics collected in national surveys for general purposes. They should only be used to indicate the broad variation of the disease across geographical space. However, these data show in general terms that Uganda, Rwanda, Tanzania, Congo, Zaïre, Zambia and Malawi are probably among the worst affected countries in Africa. It is likely that these countries are situated nearer to the source of the epidemic in Africa and are therefore a number of years ahead in the development of the epidemic. This statement in no way implies confirmation that the disease originated in Africa rather than in any other continent. This claim is probably unverifiable (although see Anderson, May, Boily, Garnett and Rowley, 1991, p. 582), and has caused a great deal of resentment on the part of African commentators (see Fleming, 1990; Hooper, 1990, pp. 219–23; Chirimuuta and Chirimuuta, 1987).

Cumulative cases of AIDS reported in Uganda to date are shown in Figure 2.4. The degree of under-reporting in Uganda is unknown, although we may guess at it. The AIDS Control Programme of Uganda has established a surveillance system (Berkley, Okware and Naamara, 1988), but there are unavoidable problems of the collection of anything approaching comprehensive data on AIDS-related deaths. The continuing civil unrest in the north of the country, 'reporting fatigue' and late submission or non-submission of reports from hospitals have all contributed to large under-estimates. All we can conclude from Figure 2.4 is that there has been an uninterrupted rise in the numbers of deaths from AIDS over the period of report.

A national serosurvey was carried out in Uganda in 1987–8 but partial results have only recently been released by the government. In late 1989, the government newspaper, *New Vision*, reported that at the time of the serosurvey there were 790,522 cases of HIV infection. It has been reliably estimated that the national total must be in the order of 1.3 million at the present time (Uganda AIDS Control Programme, reported in *WorldAIDS*, March 1991, p. 3). The implications of these levels of seroprevalence for future mortality have been examined along with other African countries

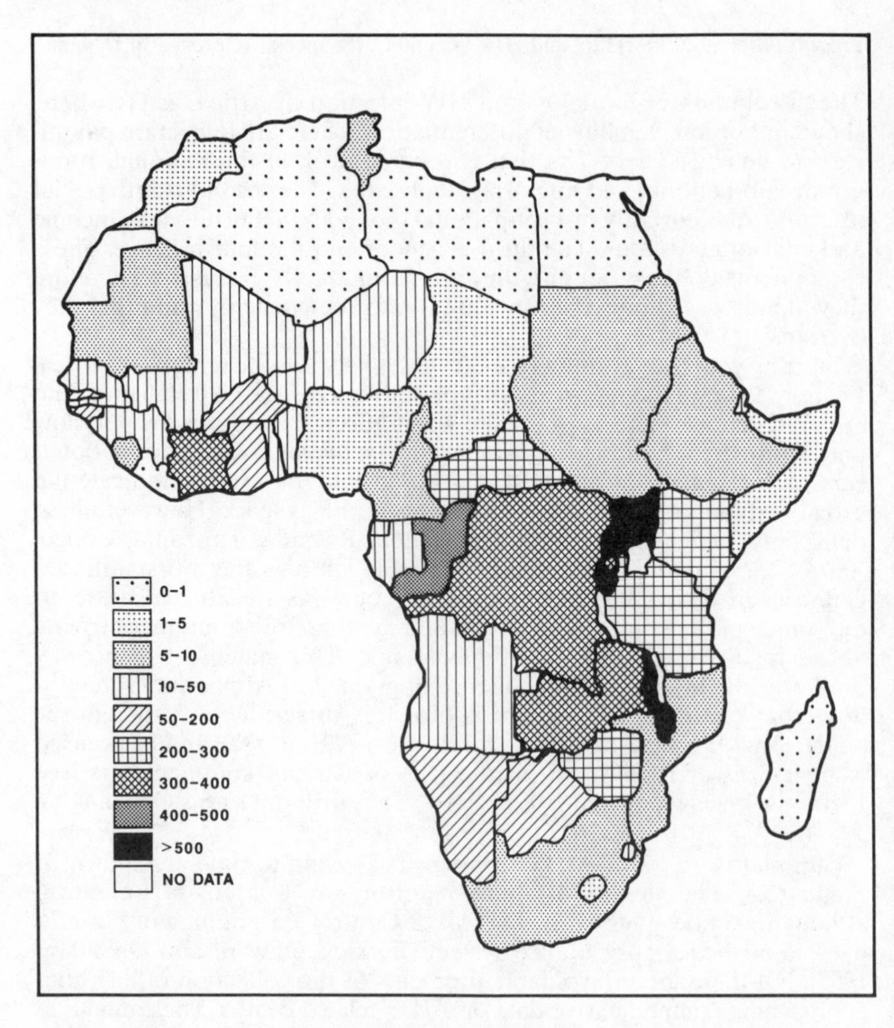

Figure 2.3 AIDS cases per million in Africa, 1990

Sources: WHO Epidemiological Record; WorldAIDS, 1990 and 1991.

in the demographic models discussed above. Early in 1991, the President of Uganda, Yoweri Museveni, was informed of a scenario derived from a version of the model developed by Bongaarts (1988), which indicated that the impact of AIDS could reduce population estimates for the year 2010 from 37 million to 20 million (*WorldAIDS*, March 1991). The finding that even those countries of the Pattern II type most severely affected by AIDS will not suffer absolute population decline may surprise the intuitively reached conclusions of many. But other models (Chin and Mann, 1989;

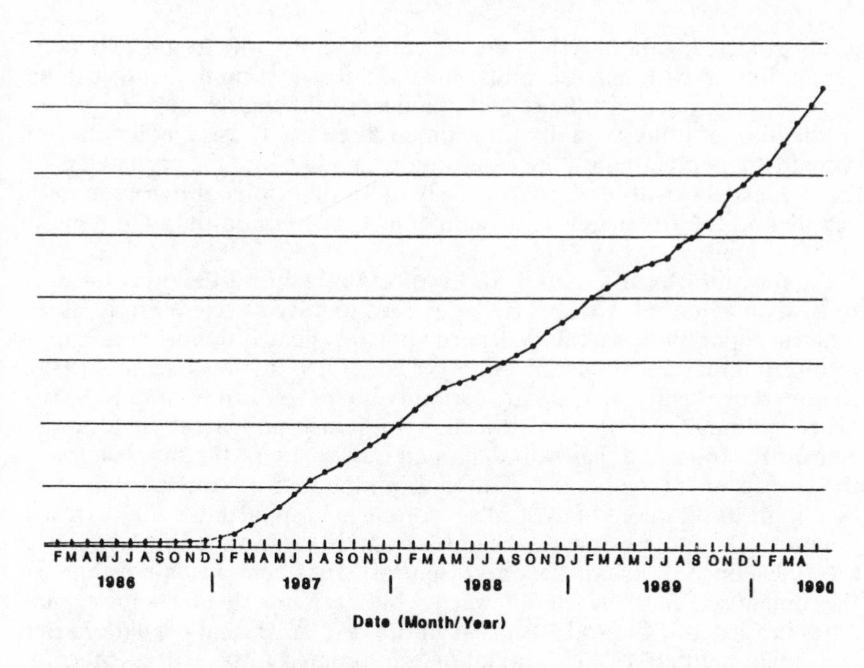

Figure 2.4 Cumulative AIDS cases in Uganda up to May 1990 (thousands)

Source: AIDS Control Programme, Entebbe, Uganda.

Anderson, May and McLean, 1988) also arrive at the same conclusion, although Chin and Mann caution that if infection increases in urban areas and spreads extensively to rural areas, then the potential for a negative population growth rate will be present. Certainly, the most recent view of one group which is researching these issues tends towards pessimism, suggesting that 'in the absence of major changes in behaviour or the development and effective distribution of better drugs or a vaccine, AIDS is likely to induce significant demographic changes in some African countries. The fact that the disease already appears to be the leading cause of adult mortality in certain urban areas in Africa lends support to this conclusion' (Anderson, May, Boily, Garnett, Rowley, 1991, p. 588).

Unfortunately, both these developments of the pandemic seem to be taking place in Uganda (Larson, 1989, as well as the detailed evidence presented in this chapter). Gregson (1990) has run the model developed by Bongaarts using Ugandan data. His findings corroborated the prediction that population growth would not be negative, under the assumptions of the model. The other predictions were: HIV seroprevalence will continue to rise sharply to peak at around 16 per cent in the late 1990s and then gradually stabilize around 10 per cent; crude death rates will double from levels expected without AIDS; the dependency ratio (of parents to dependent children) will increase; and a substantial reduction in deaths can be achieved by changes in the sexual practices of the highly mobile and the eradication of other sexually transmitted diseases. These predictions are crucially dependent upon the assumptions made about the parameters of the model. Many of these (particularly the proportions of highly mobile people and the frequency of sexual contact, to mention but two) remain open to doubt.

We turn now to differential AIDS infection within different countries. In East and Central Africa, AIDS appeared to have affected urban rather than rural populations; this was particularly indicated by earlier research results (Carael et al., 1985; Amat et al., 1989; Lavreys et al., 1989). Reported deaths from AIDS are certainly higher for urban than for rural areas, although new foci of infection in some rural areas of Uganda, Kenya, Rwanda and Tanzania are a characteristic of the later stages of the spread of the pandemic. Although undoubted urban bias will exist both in the proportions of ill people reporting to hospital and in the accurate diagnosis and reporting of the disease, it is clear that mortality and seroprevalence in urban East and Central Africa remain higher than in the rural areas. Very high seroprevalence has been reported for some capital cities in East and Central Africa, as in the case of Kampala where 25 per cent to 30 per cent of the population is estimated to be HIV+ (Merson, 1991, p. 50). Table 2.2 shows the estimated seroprevalence for urban and rural populations in selected countries of Africa. Note that these statistics are not often derived from large or representative samples and that they are at least four years out of date at the time of writing.

Part of the higher rates of infection in urban areas can be explained by the concentration of people from particular socioeconomic groups living there. The hypothesis is that higher socioeconomic status is associated with more frequent sexual partner change and the ability to travel to other urban centres (of infection). There is considerable evidence that higher HIV infection rates are found in those with higher socioeconomic status and education (Traore, 1991a, p. 47). For example, in a Rwandan sample almost a third of HIV positives in the sample had been through secondary or higher education, while in a Zambian sample, 33 per cent had spent 14 years or more in education and 24 per cent between 10 and 14 years (Traore, 1991a, p. 47). Local evidence in Uganda discussed below clearly supports the existence of this trend. The implications of these extra high predicted rates of mortality for the educated élites will be discussed in more detail in Chapter 9. The press has highlighted these. For example, The Sunday Times (London, 1.7.1990) states: 'AIDS cuts deadly swathe through African leaders', and goes on to identify politicians, professors and police chiefs succumbing to the virus in the Central African Republic. Michael Merson, the Director of the WHO Global Programme on AIDS, has stated that in some African countries 10 per cent to 15 per cent of the middle class – the industrial workers, teachers, army personnel and political leaders – will die in the 1990s (Traore, 1991b, pp. 50–2).

Table 2.2 Estimated seroprevalence for adults (15–49 years) for cities and rural areas in selected African countries, circa 1987

	HIV seroprevalence (%)		
Country	Cities	Rural	Population ('000) infected with HIV
Uganda	24.1	12.3	894.3
Rwanda	20.1	2.2	81.5
Zambia	17.2	.	205.2
Congo	10.2	.	46.5
Cote d'Ivoire	10.0	1.3	183.0
Malawi	9.5	4.2	142.5
Cent. Afr. Rep.	7.8	3.7	54.3
Zaïre	7.1	0.5	281.8
Ghana	4.7	.	98.7
Burundi	4.3	.	15.0
Tanzania	3.6	0.7	96.6
Zimbabwe	3.2	0.0	30.9
Kenya	2.7	0.2	44.5
Cameroon	1.1	0.6	33.2
Mozambique	1.0	0.6	43.5
Sudan	0.3	.	6.8
Nigeria	0.1	0.0	8.1
Swaziland	0.0	.	0.0
Total infected persons, all African countries (including others not listed)			2,497.6
Percentage of total African population infected, 1987			0.9%

Source: Over, 1990, p. 2.

While the claim that there may be a 'hollowing out' of élites may be overplayed and does not credit the robustness of the élite cadres to replace themselves (Conant, 1988), the costs of AIDS mortality are much increased, since scarce, expensive educational resources will have been invested with a foreshortened period of economic return.

It is to be expected that any pandemic will develop though space and time and therefore there will be important variations of levels of infection and mortality from one area to another at any one point in time (Thomas, 1990, pp. 5–6; Raynham-Small and Cliffe, 1990). Movement patterns of infected people therefore become important conduits of infection from one place to another. The circulation of African élites between the capitals of Africa, and at an earlier point of the pandemic between the United States and Africa, has been claimed to have been one pattern of movement responsible for the early diffusion of AIDS (Conant, 1988). The migrations of labourers both for agricultural work and mining (Parker, 1991) are others. The circulation of prostitutes between cities and between the city and their homes in rural areas also seems to have been a factor in urban-urban and urban-rural spread. Truck drivers have been cited as significant vectors of the spread of the disease, particularly along the highway linking Nairobi through Kampala and west towards Kigali (Brown, 1990). Maps showing the rising levels of seroprevalence along this route, especially by the northern shores of Lake Victoria (Okware, 1987) bear witness to this mode of spread. Micro-level data from Rakai District also shows much higher seroprevalence at roadside locations, significant at the <.01 level (Serawadda et al., 1990).

There is one data set for mortality in Rakai District, Uganda, which illustrates well the nature of micro-regional variations in mortality. It is also one in which it is possible to place a reasonable degree of trust. The data are derived from a 100 per cent census of the number of parents dying from all causes since 1971. This census was organized under the auspices of the Save the Children Fund and designed, supervised and analysed by Dr Susan Hunter of Makerere University (Hunter and Dunn, 1989; Dunn, Hunter, Nabongo, Ssekiwanuka, 1991). Its aim was to identify the number of orphans in two districts in Uganda (Hoima and Rakai). Note that an 'orphan' in Uganda is defined as a child under 18 years of age with either one *or* both parents dead or missing. This definition inflates the estimate and should be borne in mind by readers who hold the definition of an orphan as a person who has lost both parents. Figure 2.5 shows the year of death of all parents of 'orphans' in both districts.

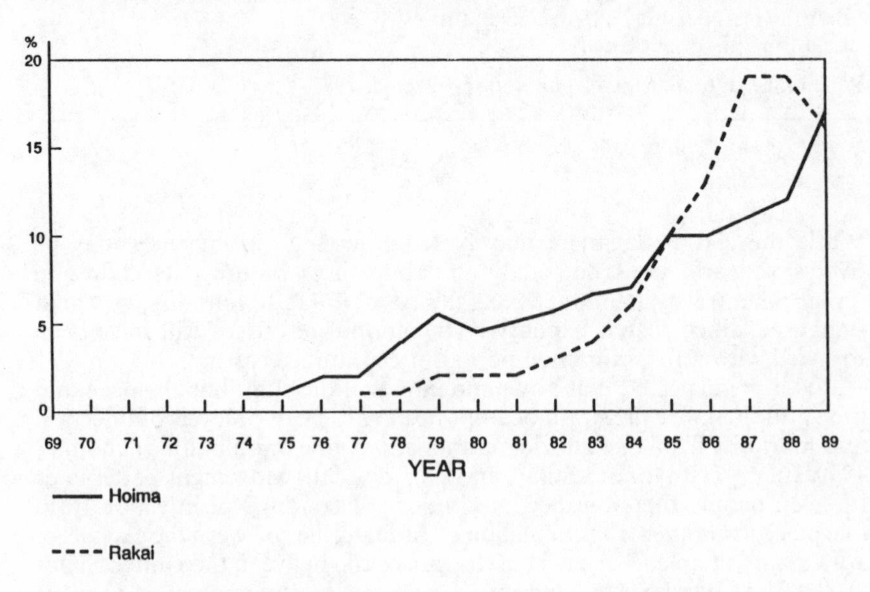

Figure 2.5 Year of death and number of parents of 'orphans' in Rakai and Hoima Districts, Uganda.

Source: Hunter and Dunn, 1989; Susan Hunter, personal communication.

The decline in deaths in Rakai District for 1989 was probably due to the fact that the survey was carried out before the end of that year. It can be seen that there is a rapid rise in the percentage of parents' deaths in Rakai District after 1983. Hoima District was severely affected by war in 1978–80, but the rise in mortality during the mid and late 1980s is also probably due to the AIDS pandemic reaching the district some five years or so after it first appeared in Rakai.

Some more disaggregated data are reproduced here for nine sub-counties from Rakai District. Figure 2.6 shows the location of the sample sub-counties and Figure 2.7 shows the cumulative deaths of parents through time for each sub-county. It should be noted that these statistics are uncorrected for population growth. Even with a constant death rate it would be expected that an increasing number of parents die in later years for reasons of population growth alone, but the lack of reliable data on the population as a whole makes correction for this factor impracticable. However, the rapid increase in mortality in some sub-counties is very marked. In the sub-counties of Kakuuto, Kalisizo and Kyebe, mortality increased greatly from about 1982, which is approximately five years ahead of the increase in mortality for the country as a whole (see Figure 2.4 for comparison). For example, the sub-county of Lyantonde is an intermediate case between sub-counties such as Lwamagwa or Byakabanda and those mentioned earlier. In the worst affected sub-counties, in a population of about 30 thousand, nearly 2,000 parents had already died in 1989–90. Taken together, the age profile of fatalities (not reproduced here) and the form of the curve both strongly suggest that AIDS is an increasing contributor to mortality in these sub-counties. Further discussion of the orphans themselves is reserved for Chapter 7.

Another important observation from these data is that the spatial pattern of the diffusion of the pandemic as it develops through time means that generalization from seroprevalence or mortality at one point in time or space to others will simply be erroneous. As the pandemic develops, the experience of mortality in any one region will be patchy.

Micro-regional variations in seroprevalence can similarly be expected to occur and to give some insights into the spatial epidemiology of the disease. A serosurvey was undertaken by a team from Columbia University with counterparts from Ugandan institutions during 1989; some of their preliminary findings were published in poster form at the IVth International Conference on AIDS and Associated Cancers in Africa in the autumn of 1989 (Musagara et al., 1989). Tables 2.3a and 2.3b show age and gender specific seropositivity rates for Rakai Districts in South-Western Uganda (which was the main field site for the present project). Table 2.3a shows overall seropositivity by age and gender for the district as a whole; Table 2.3b differentiates between seroprevalence in trading centres and rural areas.

These rates of seroprevalence are, as far as we know, among the highest recorded for a general population living in a rural area anywhere in Africa. They indicate the likelihood of a marked increase in future mortality

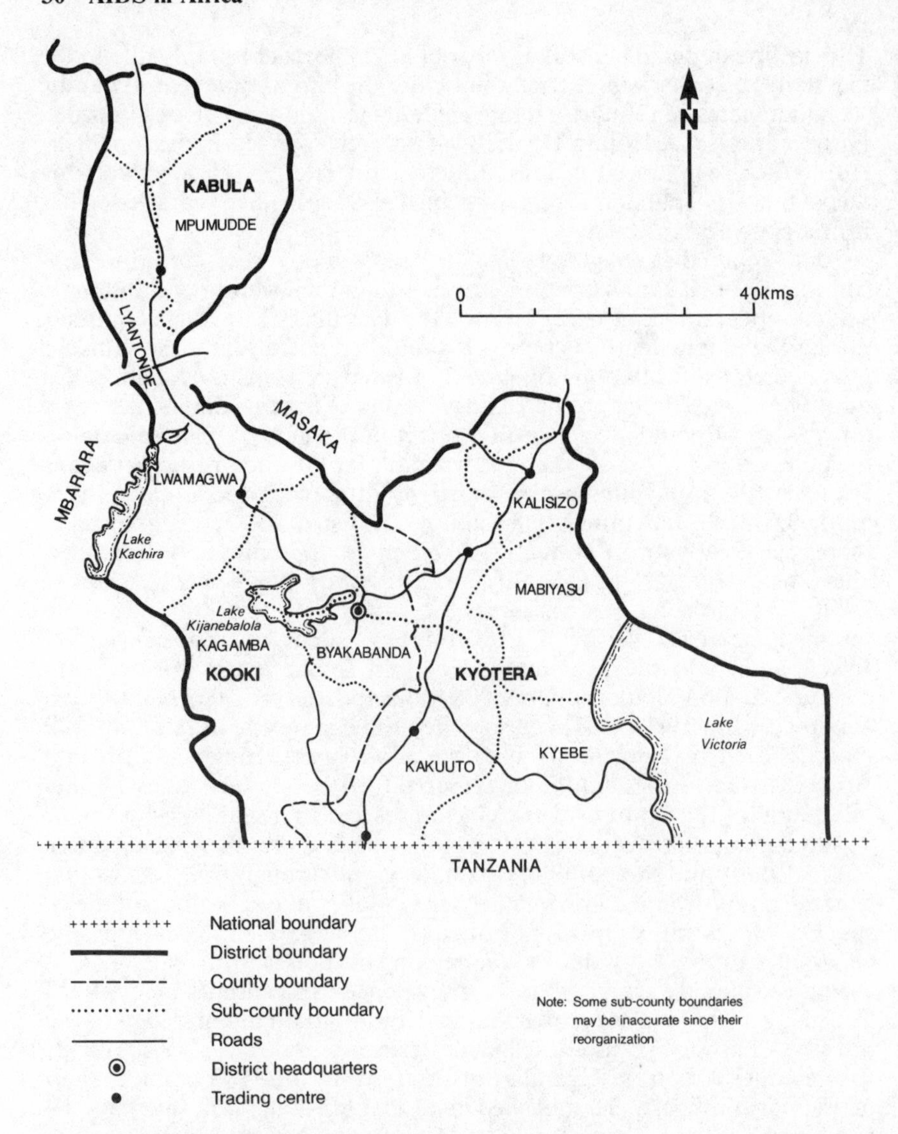

Figure 2.6 Sub-counties, towns and roads, the Rakai District, Uganda

in this part of Uganda. Given what we know of the progress of the disease in Africa, such seroprevalence will be matched by a corresponding number of deaths in future, probably within five years from initial infection (Fleming, 1990). However, unless the distribution of the date of infection of the population is known or can be estimated, it is impossible to project the numbers progressing from infection to full-blown AIDS each year.

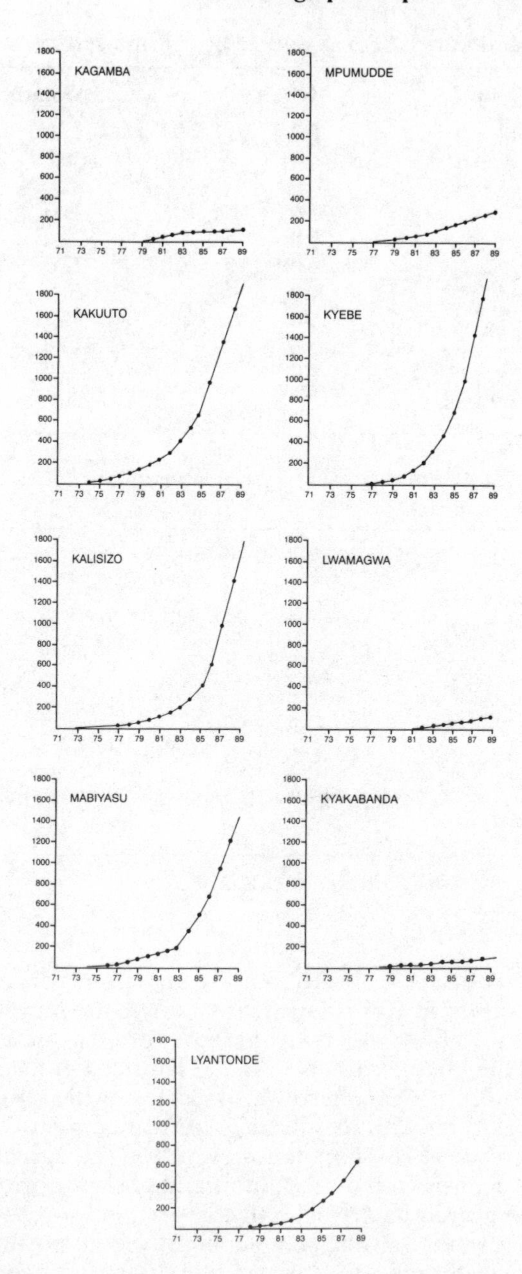

Figure 2.7 Cumulative numbers of deaths of parents by sub-county, Rakai District, Uganda.

Source: Susan Hunter, personal communication 1990.

Table 2.3a Rakai District: Seroprevalence by age and gender (percentages)

Age Group	Men	Women
10 - 14	0.0	5.0
15 - 19	5.0	30.5
20 - 24	25.0	42.0
25 - 29	35.0	31.0
30 - 34	30.0	22.0
35 - 39	15.0	22.0
40 - 44	20.0	20.0
45 - 49	19.0	15.0
50 - 54	5.0	14.0
55 - 59	0.0	9.0
> 60	0.0	0.5

Table 2.3b Rakai District: Seroprevalence in trading centres and rural areas (percentages)

Age Group	13–19	20–29	30–39	0–49	50–59	>60	Unknown	Total
Trading Centres*								
Male	3	43	30	29	8	0	0	25
Female	35	52	35	22	25	8	0	38
Rural								
Male	2	19	19	10	7	0	12	11
Female	11	25	10	18	9	0	6	14

*In Rakai a trading centre is a small settlement with perhaps half a dozen shops and a bar or two.

Source for these two tables: Musagara et al., 1989.

Serawadda et al. (1990) used multiple and simple regression on data from a large survey in Rakai District to analyse the significance of some independent variables which could have been associated with seroprevalence of the sample clusters in Rakai District. For men in the sample, the most significant characteristics associated with seropositivity were negative correlations with the distance of their community from a main road, the percentage of population involved in agriculture and the percentage of houses with a mud floor; and positive correlation with the percentage employed as drivers and the percentage with some primary education. For women in the sample, negative and positive correlations were the same as for men, with the addition of positive correlations associated with the presence of a bar or hotel in the community and the percentage of the population working there.

These statistics corroborate many of the hypotheses concerning differential infection elsewhere in Africa. As previously noted, infection has predominantly affected those with a higher level of education and with more expensively constructed houses. Infection has taken place at roadside bars and hotels often situated in trading centres (which, incidentally, were famous as far away as Kampala in the days before the pandemic made itself apparent and trade fell away; Hooper, 1990). Community access to roads and bars has particularly affected female infection rates. The survey showed that although these variables varied in the expected direction, as for men, they were not significant where women were concerned (Serawadda et al., 1990). This suggests that local women, who had access to employment opportunities in bars near to their place of residence, became infected by and in turn infected their customers, who had predominantly travelled from outside the locality and were of higher economic and educational status. More remote communities showed a lower level of infection, which supports evidence of the lower level of mortality among parents in the more remote sub-counties, found in the orphan census discussed above (Hunter, 1989a, 1989b, 1990a; Hunter and Dunn, 1989; Dunn et al., 1991). The social and economic circumstances behind these simple correlations are discussed in the next chapter.

There remains to be discussed the gender component in differential infection and mortality from AIDS. From data supplied by the AIDS Control Programme of Uganda, it can be seen from Figure 2.8 that women are infected and die at a younger age than men.

The average age of AIDS patients for all Uganda is 32 years for men and 27 for women. The ratio of male to female mortality in Africa as a whole has long been assumed to be around parity as a characteristic of Pattern II-type infection. In fact, considerable variation has been reported in the male-to-female case rate - 1.1:1 in Zaïre (Piot et al., 1984). 1.9:1 in Rwanda (Van de Perre et al., 1984). Most recent data (Berkley et al., 1990; Chin, 1990; Fleming, 1991) suggest it is possible that in general women are infected more than men in central Africa. Ugandan findings suggest that women may now be more likely to be seropositive than males, at a relative risk of 1.42 (Berkley et al., 1989); this is supported by data presented above in Tables 2.3a and 2.3b.

On the other hand, our own surveys in Rakai District, corroborated by Hunter's work, indicate that there is, and has been over the past eight years or so, a much higher mortality of men (fathers) than women (mothers). The reason for this remains unclear, although a number of hypotheses suggest themselves. One is that the greater differential mortality of men at the present time reflects earlier infection by men from small foci of high-risk women (bar-tenders, beer-sellers and prostitutes at trading centres), who then passed the infection to their wives and to other local women near their homes at a later date. If this hypothesis holds, then we may expect a corresponding increase in female mortality rates in the future. An alternative hypothesis is simply that while men and women have similar cumulative death statistics, the reporting of men to health facilities and of men's deaths are both higher. Both of these hypotheses currently remain speculative.

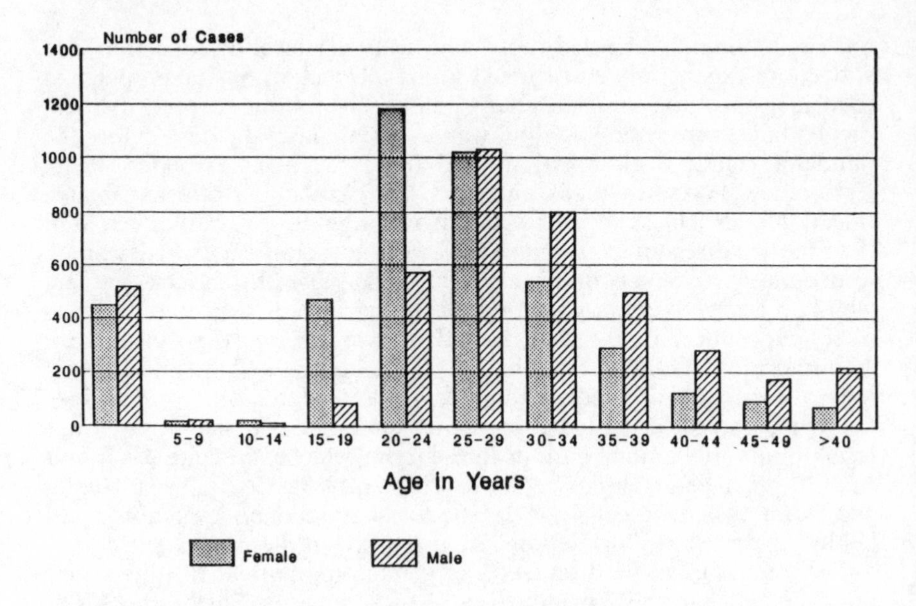

Figure 2.8 AIDS cases by age and gender, Uganda 1989

Source: AIDS Control Programme, Entebbe, Uganda, 1989.

Socioeconomic implications of AIDS-related mortality

The social and economic implications of the present and future levels of mortality principally involve the loss of people as family members and as producers. Figure 2.8 suggests that the highest death rates occur in the 20–29 year age cohort for women and the 25–34 year age cohort for men. These are the years when people are most likely to start becoming parents. In Uganda, the age-specific fertility rates for women in the relevant age cohorts are as follows:

Table 2.4 Age-specific fertility for three age cohorts in Uganda

Age cohort		Rate
20–24	=	0.33
25–29	=	0.32
30–34	=	0.27

Source: Government of Uganda, Ministry of Health, Demographic Health Survey, 1988–9.

Table 2.6 shows that, in the villages surveyed, for women the most seriously depleted cohort is aged 25–29, while for men it is the cohort aged 30–34. The death of one or both parents creates large numbers of 'orphans'. The enumeration of one or two-parent orphans in Rakai District undertaken by Resistance Committees (RCs) for Hunter's survey showed that in a total population of about 354,000, there were around 23,500 such orphans, or 12.6 per cent of the total of all children under 15 years (Hunter and Dunn, 1989). Table 2.5 shows the greatly increased proportion of orphans in two sub-counties of Rakai District, which have particularly high mortality of parents, as compared with the expected percentage of orphans from a stable population with similar characteristics other than the presence of AIDS. It is worth noting that these statistics were collected in 1989. In a television programme on this subject about this area shown in the United Kingdom during 1991, the total number of orphans was given as 40,000, but it has not been possible to verify this estimate. However, given the seroprevalence in the area and the dates of the earliest outbreaks, it is probable that, nearly two years after the original census was undertaken, a marked increase in the number of orphans has occurred.

There is some anecdotal but unsubstantiated evidence from our own research and from other studies that, even if the orphan survives its seropositive mother, the ensuing death of one or both of its parents precipitates a crisis of caring and parenting, resulting in the raised mortality of orphans.

Finally, younger adult household members perform the important role of caring for the sick and elderly. If they themselves are sick or have died, this important role can no longer be performed. Younger adults often participate in the management of their household. They make management decisions about earning cash, expenditure on consumption, education and other expenses, the provision of an adequate diet and so on. The death or sickness of these members removes essential and often unremarked skills from the household.

A random sample of people interviewed by the authors and their colleagues in Rakai and Kigesi Districts showed that there was already

Table 2.5 Proportion of orphans in two sub-counties of Rakai District

	Proportion of orphans reported in each age group		**Proportion in a stable population like Uganda**
Age group	**Sub-county 1**	**Sub-county 2**	
<1	3%	1%	8%
0–4	15%	16%	22%
5–9	39%	40%	33%
10–14	42%	43%	32%

Source: Hunter and Dunn, 1989.

Table 2.6 Comparison of age structure between two sets of villages, males and females, Rakai and Kigesi Districts, Uganda, 1989

MALES

Age groups	VILLAGES 1 & 2 number	VILLAGES 1 & 2 percent	VILLAGES 3, 4, 5 number	VILLAGES 3, 4, 5 percent
0–4	21	16.8	53	12.7
5–9	18	14.4	58	13.9
10–14	16	12.8	52	12.5
15–19	15	12.0	51	12.3
20–24	18	14.4	38	9.2
25–29	11	8.8	38	9.2
30–34	**1**	**0.8**	**30**	**7.2**
35–39	4	3.2	20	4.8
40–44	7	5.6	18	4.3
45–49	3	2.4	16	3.8
50–54	6	4.8	13	3.1
55–59	0	0.0	4	0.9
>60	5	4.0	24	5.7
Total	125	100.0	415	99.6

FEMALES

Age group	VILLAGES 1 & 2 number	VILLAGES 1 & 2 percent	VILLAGES 3, 4, 5 number	VILLAGES 3, 4, 5 percent
0–4	21	13.8	47	11.5
5–9	19	12.5	59	14.5
10–14	27	17.7	48	11.7
15–19	27	17.7	53	12.9
20–24	12	7.9	46	11.3
25–29	**4**	**2.6**	**38**	**9.3**
30–34	10	6.6	25	6.1
35–39	10	6.6	29	7.1
40–44	6	3.9	16	3.9
45–49	7	4.6	13	3.2
50–54	2	1.3	10	2.5
55–59	6	3.9	5	1.2
>60	1	0.6	19	4.6
Total	152	99.7	408	99.8

Source: Authors' own fieldwork.

evidence of unusual demographic change in the worst affected areas. These data indicate that the age cohorts that are the most active as parents and producers are apparently suffering increased mortality rates. The sample size is too small to draw conclusive and statistically significant results. However, it may be taken as indicative of what may be happening in areas of high levels of infection in Uganda and elsewhere. The age and gender distribution of the populations of these communities is shown in Table 2.6. Villages 1 and 2 in this table were known to have very high infection rates and to have suffered exceptional mortality, while villages 3, 4 and 5, although not free from reports of deaths from AIDS, were situated further from foci of infection. The bold figures in the Table indicate a comparative under-representation of that age cohort in villages 1 and 2, as compared with that in villages 3,4 and 5.

Thus Table 2.6 dramatically shows the impact of AIDS upon the most productive age groups in these communities. Note the depleted cohorts of men in the 30 to 34 and 35 to 39 age range and of women in the 20 to 24 and 25 to 29 age range. These coincide strikingly with the cohorts five years younger which show the highest seropositivity rates in Tables 2.3a and 2.3b above, the average survival rate from infection to death in Africa being about five years. For comparative purposes, we can assume that in a stable population in a country like Uganda, we would expect that about 5.5 per cent of males would be between 30 and 34 years (actual percentage in the AIDS-affected villages 1 and 2 was only 0.8 per cent). Similarly, about 7.5 per cent of women in a stable population should be between 25 and 29 years of age, whereas the observed percentage in the AIDS-affected villages was only 2.6 per cent. There are quite good grounds for stating that there are some clear, pronounced differences between these two sets of villages, most probably due to the impact of AIDS. However, it is possible that they reflect factors other than AIDS-related mortality. For example, it could be argued that the data reflect differential labour migration and marriage patterns in the two sets of villages. However, these data refer to *families* and not households; absent children of the household who were living away from the parental home are included. There are bound to be problems with recall and recording of absent family members. This is further complicated by the practice of serial polygyny on the part of many men in this area, but such errors should be equally distributed between the two sets of villages. It is also possible that the respondents in villages 1 and 2 had more difficulty in recalling their absent members for some reason. This is a possible explanation, given that in both of these villages young men leave for varying periods to work as fishermen in the lakeside settlements, and thus do not fall easily into either the category of an absent member or a household member: they are something in between. However, this does not seem an adequate explanation for the observed pattern and we consider that real AIDS-related deaths in the specific age cohorts are a more likely explanation of the differences between villages 1 and 2, and 3, 4 and 5 in Table 2.6.

Summary

There remains much uncertainty about the current scale and rate of growth of the AIDS pandemic in Africa. However, a number of factors has cumulatively provided evidence about orders of magnitude. It is possible to state that AIDS-related mortality will rise rapidly in some areas of Africa and will constitute the main cause of death for children and adults in some of the age cohorts between 15 and 50 years. There are important components of differential infection and mortality. These are due to the spatial diffusion of the disease and can be noted from the national, regional, micro-regional and rural-urban perspectives. Gender, age, socio-economic status and occupation are other important components of the differential rates of infection and mortality.

Note

[1] **seroprevalence** – the rate of HIV infection in a population, usually given as the percentage of a sample of people from whom blood has been taken and who have been found to be HIV+, or, in the case of a whole country, as the number of people who are HIV+ per million of the whole population.

A note on further reading

The literature on the subject matter of this chapter is, inevitably, very technical and quite specialized. Key articles are: Anderson, May and McClean, 1988; Bongaarts, 1988 and Anderson, May, Boily, Garnett and Rowley, 1991.

3

Coping with AIDS – Rationality, Explanation and Action

We use the term coping as a general term to include defense mechanisms, active ways of solving problems and methods for handling stress. (Murphy and Moriarty, 1976)

Households make pre-emptive decisions following a drought in an attempt to mitigate the somewhat predictable effects of a severe (food) shortage or market distortions perhaps six months distant.

Most response strategies to actual or potential food shortage are in fact extensions of practices conducted in some measure during a normal year. The vast majority are *in situ* and can be broadly classified as self-help. (Watts, 1988)

Human beings are rational creatures. However, their rationality is only applied to problems in the context of the knowledge and beliefs they have. Coping is about the ways in which we all recognize that our normal expectations of how life is and ought to be are adjusted when we realize that 'normality' has, for whatever reasons, switched to 'abnormality'. In recognizing that such a transition has occurred, we search for explanations of the new circumstances in which we find ourselves; we adjust our expectations and we search for courses of action that will enable us to achieve whatever goals are culturally significant for us. These processes require us to develop not only practical strategies for coping, but also language and concepts to deal with the new situation. In this chapter and the next, we explore some of these conceptual and practical transitions as they are occurring in communities in Uganda that are confronting the intense problems presented to them by the AIDS pandemic.

Crisis events arise from the impact of a hazard on human lives. They occur from time to time in individual lives and in the lives of whole communities and societies. Such events call for the mobilization of emotional, intellectual, human and material resources to cope with their impacts. When people know that an event will occur because it has happened in the past for them or for other members of their community, they develop ways of coping with it in advance. Coping strategies require that links exist between the behaviour of individuals and of groups such that an entire community or society may mobilize its resources to confront the hazard as effectively as possible within the limits of existing and accessible technology and forms of social organization. Depending upon the social and cultural context, information about such events becomes 'knowledge' which facilitates such mobilization. This knowledge may reside in 'tradition', in 'custom' or in 'science', and may facilitate 'problem solving' or the production of 'policy'.

In order that coping is possible, there are certain prerequisites. First of all, the hazard event itself must be socially perceived and recognized as following a familiar pattern or as being a novel event that can be related to pre-existing experience. In other words, it must be coded into the existing cultural discourses, language and sets of meanings characteristic of the affected society. The possibility of coping with a crisis may be experienced by individuals and by human communities in a variety of ways: as feasible, because the event is not entirely novel or unexpected; as difficult because it falls at the extreme of past experience; or as impossible because it is so novel that existing concepts, rules and techniques offer no guide to coping in the new circumstances. For coping to be possible, and central to this process of coding such hazard events, it is necessary, at both the individual and social levels, that information about the event can be obtained and processed into a usable form within the affected human group (Douglas, 1985). A crisis must be identified and understood in a way that means something to the affected people before it can be coped with.

The second prerequisite for coping is that people make the assumption that the basis for decisions in the social, economic and natural environments will not have changed and that other people and the natural environment will behave in familiar ways. This assumption is a necessary part of the response to 'normal' crises. It may also be the Achilles heel of any response to the 'abnormal', because some unprecedented crises, like the AIDS pandemic, occur when the past does not provide a satisfactory guide to the future. The assumptions about the continuity and stability of the decision environment rests on a bounded rationality and a view of the universe of crisis events, which assumes that the general features of hazard impact are understood, coded and therefore manageable. However, in the case of the AIDS epidemic, the bounded rationality derived from past experience of drought, famine, war and social and economic disasters – events all too familiar to the people of Uganda – does not provide adequate guidance for coping mechanisms either for people in small rural communities or for officials responsible for government policy. New processes of understanding and of responding to a new type of hazard are required.

In 'normal' times people live their lives in the knowledge that, sooner or later, particular crisis events or disasters will occur. This is the knowledge of expected risks in relation to which people have some stored subjective estimates and experience of how to cope. This experience is coded into the rites of passage, ceremonial, moral codes and the very structure of societies, as in the rules determining who should take responsibility for an orphan on the death of its parents or for a parent in old age, or of how to deal with crop failure or cattle disease. These are problems common to all societies and are solved differently depending on culture, societal scale and level of technology. For example, births, deaths and marriages will occur for all households at some time. Similarly, droughts, floods or other disasters will inevitably cause widespread loss in the future for those living in a physically hazardous environment. Also, people do not like conditions of uncertainty where there are no known ways of coping with a particular disaster through accepted systems of rights and obligations between individuals in local society or provided through the bureaucracy of the state. All of these provide safety nets and support groups. The unprecedented event, such as the AIDS pandemic, creates a situation of uncertainty.

In Uganda, where some communities and households are increasingly affected by illness and death as a result of AIDS, the balance is perceived over time to move from the known towards the unknown, from the normal to the abnormal, and from knowledge that is coded into the moral structure of society to new experience that has to be coded and processed into that particular form of information defined as 'knowledge'. At the margin, another single death may seem like any other, and, of course, all societies have well-established mechanisms to cope with a death. But when large numbers die and are perceived to do so as a result of unprecedented causes, established coping mechanisms and taken-for-granted knowledge may begin to show signs of stress and inadequacy in the face of a welter of new, unpleasant experiences. In such circumstances, expectations derived from past convention increasingly become a poor guide to the solution of current management problems. Coping becomes difficult. The conditions of everyday life become increasingly uncertain and therefore stressful and puzzling. Confronted by such uncertainty in the decision environment, it is to be expected that a range of 'experiments' will be undertaken by households and communities as their experience of the new situation increases. Such 'experiments' can be seen as attempts to reverse the balance between the known and the unknown in the search for normalization. One of the first steps in coping with a crisis is to explain it.

From a rational perspective, the most effective way of coping with the AIDS pandemic is to change sexual behaviour and also to avoid possible sources of contamination such as infected hypodermic syringes or implements used in ritual scarification. In order to do this, it is necessary to have an accurate explanation of how the disease is transmitted. Information has to be translated into knowledge and then into action (principally in this area involving a change in sexual behaviour). However, this is not an adequate account of how real people actually deal with these risks. In reality, most of us cope with risks through a combination of rational and non-rational responses. This is not to suggest that we act 'irrationally', but rather to draw attention to the fact that there is a considerable difference between 'non-rationality' and 'irrationality'. The distinction is as follows. Rational behaviour has a clear goal where resources are used efficiently in order to achieve it. One of the resources is adequate information and knowledge about the hazard and about the risks attached to that hazard. The claim of scientific explanation to be in some sense 'true' and thus superior to other forms of explanation rests on the belief that experimental evidence provides the best account of the factors in a causal chain that leads to the appearance of a particular phenomenon.

The scientific explanation of AIDS rests on establishing a chain of causality between the symptoms of illness and the ways in which the virus compromises the human immune system. Rational response to this knowledge involves breaking this chain of causality at some point – within the body (impossible at the present time as there is no cure) or between bodies. When people do not act in accord with this line of reasoning, it is not necessarily the case that they are acting irrationally and denying the validity of the scientifically established chain of causality. They may be acting in accordance with another explanation, based on different assumptions about the nature of the causal chain. These assumptions may use the language of 'chance', 'luck', 'witchcraft', 'sorcery', 'sin', 'morality' or 'punishment for moral misdemeanour'. There are many examples of such thinking, where the language and concepts used in relation to the disease appear to a present-day observer to bear little direct relation to the issue at hand.

A number of cases illustrate the range of different rationalities. For example, Davenport-Hines (1990) tells us how sexually transmitted diseases have been treated as a moral rather than a medical issue in Britain. Gluckman (1955) shows how malarial illness may be treated as an issue of witchcraft, despite knowledge of the roles of the disease vectors. Lyons, (1988a, 1988b) writing on sleeping sickness in Africa, shows us how that disease was used in the context of colonial government as a rationale for the relocation of the population. Vaughan (1991) provides a case study of the ways in which the languages of medicine, morality and colonial administration were mixed together, resulting in the production of policies that eventually dealt not with 'medical' problems but with the social and political prejudices of the powerful. Packard's history of tuberculosis in South Africa (1989) shows how the way in which it was discussed and treated often had more to do with political and social prejudices than with medical science.

The point is that none of these explanations and responses to disease can be considered as 'irrational' as long as the assumptions (however misinformed they may appear with hindsight or from a different cultural and historical perspective) are used as the basis for future action and the logic of those actions flows from the assumptions. These other forms of explanation (and of scientific explanations, too) have a 'bounded rationality'. They take into account certain limited knowledge and information in the decision process and give weight to some information or knowledge that others may weight differently. Some of this difference in weighting may derive from 'custom', 'superstition', 'belief', 'values' or scientific method. In these circumstances, some may label this a 'non-rational' basis for the ways in which people act, but it is most certainly not 'irrational'. This term suggests that a person is acting contrary to the information or knowledge they possess. But in reality, this means that the users of the label 'non-rational' cannot or will not understand why another person or group gave greater weight to one piece of information and less weight to another – and that they would have reacted differently. In what follows, it will become clear that in Uganda, as elsewhere, people are responding rationally to the AIDS epidemic in terms of their own world view and experience.

There are wide variations in the level of awareness of the causes of the disease between different regions of Uganda and at different points in time. One of the earliest outbreaks of the disease was probably experienced in one of the small fishing villages on the western shore of Lake Victoria, which became a smuggling port in the mid-1970s. Traders involved in smuggling began to fall sick and die in 1980–1. The dominant local explanation at the time was that the traders had been involved in a major swindle of Tanzanian traders and that the latter, coming from the district of Bukoba (well-known for its powerful witches – in particular the Haya and Ziba people), had bewitched the Rakai traders and caused them to die (Hooper, 1990, pp. 40–42). The fact that the traders' wives were not long in following them to the grave was seen as confirming the potency of Bukoba witchcraft. However, another trader not involved in the original swindle also died, together with his wife. About this time public-health education was reaching school teachers and local leaders. An alternative explanation was required; the message of the health educators was assimilated but it did not completely displace the existing explanation.

Today we can find a range of syncretic explanations. It is still common to hear a composite account in which the scientific and witchcraft explanations are merged. These suggest that while the infection can be picked up during sexual intercourse, there is always a chance that it will not occur, even if the other person is known to be infected – a view not inconsistent with the facts. Thus the language of 'chance' and probability is introduced into the chain of causality. It is said that this 'chance' can be influenced by witchcraft, which, if malignly exercised, can cause the infection to pass *during that particular sexual encounter*. There are a number of other variants such as the belief of some men that, if they pay women for sexual intercourse, they propitiate one of the lesser gods, who, if not properly placated, are well-known for activating the infection. Here the languages of science, witchcraft, religion and the market are intertwined. Such an intermingling of different theories of causality reflect the ways in which individuals deal with a hazardous environment by drawing down different parts of the culturally available vocabulary to enable them to continue to behave in ways that are 'normal'. Such a process of creating a rational framework for daily life fits in with the attempt to construct normality in an abnormal and novel situation, such as we have suggested exists today in Buganda. It is a way for individuals to hold on to past normality as the decision environment changes. However, the process is not uniform. Gender, age and locality will all influence the particular combination of explanations that individuals and groups draw upon in order to explain and to cope.

In the areas of highest infection in Rakai and Masaka Districts of Uganda, for example, the medical explanation is widely held by the majority of the population and particularly by older school children. This explanation includes identification of the virus as ***buwuka*** (a small insect), which is passed between partners during sexual intercourse. It was also noticeable that in areas that had not experienced high mortality from AIDS, such as in Kabula sub-county in north-eastern Rakai District and in Kigesi District, hundreds of kilometres away near the border with Rwanda, explanations of AIDS as the result of witchcraft continue to prevail. A recent survey of knowledge, attitudes and practice regarding AIDS and AIDS prevention (Berkley, Downing and Konde-Lule, 1989) reported that 75 per cent of the sample knew the correct details of how the disease is transmitted and only 4 per cent ascribed it to witchcraft. This is broadly in agreement with our own findings, except that our extended discussions revealed that many composite (and complex) explanations continue to survive.

Even in those parts of south-eastern Rakai District that have experienced the highest mortality (see Chapter 2), there is much evidence, at least from men, of continuing risky behaviour together with its attendant rationalizations. Two quotations from a couple of men frequenting a bar in the study area illustrate this. Thus: 'If indeed AIDS exists and is caught from sex, I will be the example' (meaning 'I will be the exception to the rule'); 'People were never meant to be like timber and live forever' (meaning 'we all have to die sometime'). It is interesting that it is men who adopt these orientations. There are a number of reasons. Firstly, in the society of Buganda, as in many others, male identity is closely connected with sexual conquest and fecundity. Secondly, there appears to be little in the way of male solidarity. The disrupted history of the past 30 years has placed a premium on individual competition and survival: success in these endeavours is symbolized at least in part by sexual conquest. For these reasons, it is all the more necessary for men to keep open mental and emotional escape hatches so they can continue to act as they have in the past.

In contrast, women see the existence of the disease as a threat to themselves and their children, but one against which there is little defence given their subjection to men. Among women, one of the most distressing experiences, and a threat with which they have to live, is the birth of an infected child. Its suffering is a continual reminder of the consequences of past actions and an indication of what will certainly be in store for them, too. There is enough experience in the worst affected areas of HIV+ children who 'are born coughing' for it to be a major motive for changing sexual behaviour. And it is certainly among women that most discussion is heard of the need to change exactly that.

Two further examples show how people do not act on the medical and experiential knowledge of the infection and of the chances of catching it. The first is the belief that the act of marriage itself makes the partners immune from infection. This may have come about because both the Catholic and Anglican churches as well as Muslim leaders have urged marriage and 'zero grazing'. Although adherence to this advice may reduce future risk of infection from sexual intercourse outside marriage, it does not guarantee the seronegative status of the partners at the time of marriage nor the prevention of inter-spouse infection afterwards. The second example is the belief that a pregnant and seemingly healthy woman cannot be HIV-positive. Western readers may be reminded of similar beliefs in Europe. There is an advertisement warning of the AIDS risk in which a handsome and seemingly healthy young man boldly appraises the reader from the page of a newspaper or magazine. The caption reads: 'Before you sleep with someone, look out for signs of HIV (the virus that leads to AIDS)'. Helpful arrows point to aspects of the man in the picture such as 'healthy appetite, normal weight, perfect eyesight, clear skin' etc. (*Sunday Times*, 25 March 1990). In addition, in Buganda the idea that female beauty is a protection against AIDS infection is prevalent among men; this is mentioned by adolescent school children in their essays, discussed below.

Even if an accurate explanation of the disease and of the risks involved is accepted, current seroprevalence make even occasional sexual intercourse highly risky in the centres of infection. So occasional bouts of celibacy that may be brought on by the experience of a suffering child or the deaths of acquaintances may not constitute a real reduction in the risk of infection. The effect of advice by the AIDS Control Programme to 'Love Carefully' is not likely to reduce the risk of infection to anything like acceptable levels in the areas of most severe infection unless it implies the use of condoms without exception. Since these are rarely available outside Kampala and the largest provincial towns, and since until very recently the government of Uganda has been ambivalent in its attitude to condoms, the advice, even if taken, in unlikely to be effective. The use of condoms in Rakai District is virtually zero, although a few people reported using them in trading centres (see also corroborative evidence from Musagara et al., 1989).

A number of other strategies are perceived to reduce the risk of infection. For example, men tend to seek much younger sexual partners and wives in the belief that these women will not be infected. The result of this practice can be seen in the proportion of very young women who are now seropositive and in the differences in numbers of AIDS cases and rates of seropositivity as between men and women by age (see Figure 2.8, Chapter 2). Other men seek brides outside Rakai and Masaka Districts altogether. In some cases the husbands and wives of people who have died of AIDS have migrated, seeking a new partner elsewhere and leaving behind them the stigma of being the spouse of an AIDS victim and of living in a notorious area. These actions merely contribute to the spread of the disease to other areas.

The change of sexual behaviour to 'zero grazing' (faithful marriage) is being advised by the churches and Muslim leaders; but most religious institutions have resisted the slogan 'Love Carefully' since it seemed to condone promiscuous behaviour. In any case, there is no evidence that large numbers of people heed either counsel. The only reliable evidence as to whether or not people are changing their behaviour would be a comparison of seroprevalence at short intervals (for example, annually) in a statistically controlled population. Seroprevalence reported from Rakai District (Musagara et al, 1989) indicates that while 'zero grazing' may have been accepted at the level of answering questionnaires, it is not actually practised.

Evidence in the communities in Rakai shows there is social pressure to control circumstances that might facilitate casual sex. Rural Resistance Councils have introduced by-laws banning discos and the levels of casual sex at bars and hotels in trading centres appear to have been greatly reduced. Certainly, in the course of our research, when we stayed at small hotels and visited bars in places such as Kyotera, Kalisizo and Lyantonde, they were not the hives of activity they reportedly had been even five years ago. Transmission also occurs at weddings and funerals and some Resistance Councils have attempted to ban alcohol and to reschedule these events to daylight hours to discourage sexual encounters. There is also new legislation at the national level, which has raised the age of consent from 14 to 18 years, made homosexuality and prostitution illegal, and made the rape of a child under the age of 14 a crime punishable by death (Tebere, 1991, p. 3). Whether these responses will accelerate a change of sexual practice remains to be seen, but a useful indicator is the attitudes and knowledge of the sexually active population itself, particularly those adolescents who are about to start sexual activities and who are, for the most part, still seronegative. The focus of concern in the next section is on the beliefs and opinions of such young people: an understanding of their view of the situation is of the greatest importance for the design of policies intended to slow the spread of the disease.

School childrens' views on AIDS

In order to obtain some idea of how this age cohort is responding to the situation, 55 children aged between 14 and 17 (39 boys and 19 girls) at secondary schools[1, 2] in Rakai District were asked to write essays on themes related to AIDS. They were not briefed before the exercise. The children came from a range of ethnic groups found in Buganda – Banyankole, Banyarwanda, Banyoro and Baganda. Most of the essays were written in LuGanda (28) or English (24), while some were in a mixture of English and LuGanda (4) and one was in Lunyankole. The children were asked to write on the following three themes:

1 Mary and John are getting married. Mary's mother wants them to go to the hospital for an AIDS test.

2 Last year was good for George. He graduated from Makerere University, married Teresa and they had a baby called Peter. Now the whole family is dying of AIDS.

3 Sarah graduated last year from Makerere University. She received a scholarship for further medical training in England. Last week she died of AIDS.

These topics were chosen for the following reasons: topic one because it was observed in the course of research that many women said they would like to protect their daughters from AIDS by ensuring that both they and their intended husbands had an AIDS test before marriage. Topic two was selected because many people in Rakai believe that most of the AIDS deaths have been among the élite group – people such as professionals and business people (borne out by the serosurvey of Musagara et al., 1989). Topic three was selected specifically to gain information on the attitudes of these adolescents to women, since women in general, and particularly single and educated women, are being blamed by some men as a group responsible for the spread of the disease. The tendency of some men to scapegoat women who stand outside the structure of female subordination is an important issue in a society where, as we shall see in Chapter 5, women's subordination has been a factor in the distinctive form of the epidemic in Uganda.

Table 3.1 shows the results of a content analysis of these essays. It indicates the main themes preoccupying this group of students living in a high-impact area.

From Table 3.1 it can be seen that there are five major categories of concern among these adolescents:

1 AIDS deaths are a loss of resources to the country and to the district. This includes the idea of waste of economic investment.
2 Disapproval of behaviour of the élite who are seen as acting irresponsibly.
3 The evil intent that some men have in seeking out young girls or raping women.
4 Identification of women as the main culprits for the spread of the disease.
5 A person who gets infected is acting immorally and is dishonouring their parents.

The greatest emphasis is on the irresponsible behaviour of the educated people. The comments of these young people indicate both moral disapproval and also anger at the way in which such people are judged to be wasting scarce national resources by acting in an irresponsible manner. Their irresponsibility is also subjected to additional condemnation because of the sense that these school children have of how such behaviour indicates lack of regard for parents and ancestors. Certainly, little sympathy is expressed for the unfortunate sufferers. These adolescents

Table 3.1 School children's views of AIDS

THEME	NUMBER REPORTING
Elite deaths represent wasted money, wasted labour power and a loss to the country.	
When people discover they have AIDS they purposely spread the disease.	51
The élite should know better than to contract AIDS.	50
The death of so many élite individuals is not good for either Rakai or for Uganda	45
If people find they are seropositive, they might commit suicide.	43
Having AIDS is dishonourable to parents.	41
The gender with the largest propensity to spread AIDS is:	
women	40
men	3
neither	11
not raised as an issue.	4
Rich men with AIDS entice young girls by offering them money.	37
Collective action is needed to conquer AIDS.	20
It is important to listen to the advice of friends and experts.	20
People should have an AIDS test before they marry.	12
There is a need for mandatory testing:	
in schools	9
for everyone.	3
Women are concerned with dying:	
and leaving their children	11
childless.	6
Men with AIDS rape young girls.	5
AIDS patients should be quarantined as lepers used to be.	5
Trained people are essential and are most in need of AIDS protection.	5
Prostitutes spread AIDS.	4
Orphans are a serious issue in society.	2

Source: Data collected and translated by Dr Christine Obbo.

have accurately perceived the differential rate of infection, in which urban and élite populations were infected first and at a higher rate than rural populations.

It could be argued that some of these themes express a more general disapproval of élite behaviour which, after the disruption and turmoil of the last 20 years, does not indicate that élite members are either to be trusted or have acted with the general welfare in mind. Such comments about élite behaviour are not restricted to those who have caught AIDS. It is apparent that the educated sufferers from the disease are not afforded much sympathy, and that the perceived behaviour of members of the élite that might lead them to become infected is the object of strong moral disapproval. Thus, one essayist commented:

> The money wasted on educating people who die of AIDS is regretted by their parents. Our parents weep all the time. Students who sneak out at night to meet their lovers or who refuse to use condoms must remember their parents' sacrifices and think of their sorrow.

These essays are also of interest for what they tell us about the writers' general attitude to the question of witchcraft. In two essays on theme 2 the relatives of the dead man and woman are wealthy and have many consumer goods in their homes. They are regarded with jealousy and jeeringly referred to as ***wazungu*** (white people). After the deaths of Teresa and George in theme 2, they claimed that the two unfortunates had been bewitched. The neighbours treated the assertion with contempt – they knew that both young people had died of AIDS.

The dismissal of witchcraft in most of the stories (in all, two writers raised it as a possible cause while discounting it, five discounted it altogether and the rest of the essays did not consider it an issue at all) indicates that among this group of school children, the medical cause and nature of the disease is well understood. However, this understanding of the scientifically established chain of causation is then used as the foundation for a moral evaluation, based upon the condemnation of people for their disregard of traditional family and communal values and also for irresponsibly wasting the local and national resources that have been invested in their education. The issue is individualized; it is argued that individuals should behave responsibly and that it is individual behaviour change that will make a difference.

There is a related set of judgments that applies specifically to the behaviour of infected men. These adolescents suspect that some infected men set out deliberately to infect women, seeking out young women for sexual relations and even raping young girls for the same reason. This type of story is not unique to Buganda or to Uganda: we heard similar accounts in other places both in Africa and elsewhere (see Shilts, 1987). Thus in one version of story 1, John suspects that he may be HIV+ because his girl friend is dying of AIDS. His attempts to persuade her to elope with him in defiance of her parents fail. He resolves to leave the district, to travel to a distant place and spread the disease there as a desperate revenge for his fate. Fortunately, he has an AIDS test and is pronounced seronegative. The rationale for this view of the possible actions of AIDS victims (found in 51 of the stories) is contained in the view that: 'I did not bring the virus. Why should I die alone?' One student said that when young men found they were infected 'they were angry and wanted to destroy everyone and everything'.

There is some evidence that men do in fact behave in this way. In separate interviews with 25 men who knew they had AIDS, three respondents admitted to purposely initiating sex with women in order to infect them. In one case, rape was admitted. In 37 essays, rich men were portrayed as enticing young school girls with offers of money; in five essays they were portrayed as raping young girls. These themes reflect two very real aspects of the social context of AIDS' impact in Buganda: the tension between rich and poor and between young men without wealth and older men with wealth in their competition for sexual partners.

We have already noted that there exists a suspicion of women as major sources of infection. Thus:

> John was a promiscuous young man, but he had been fortunate and not contracted AIDS. He met Mary who had been strongly advised by her mother not to 'play sex' before marriage. She was infected and he caught the disease.

In fact, 41 essays suggested that the disease was spread by 'immoral' women. In contrast, only three essays felt that the disease was spread by 'immoral' older men visiting prostitutes and then turning to young girls. So despite the view that there were some men who purposely set out to spread the disease, it is women who are seen as the main source of infection (e.g. 'prostitutes who go with five men a day'). One writer also expressed the belief that women were more of a danger because they survived longer before the disease manifested itself. Another wrote: 'it is rumoured that it is from ladies that AIDS has spread to men in the whole world', while a young woman commented that: 'So far women are the main spreaders of the disease. Each woman who dies leaves ten men to follow her.'

A related theme in these essays concerned the dilemmas faced by the widows and widowers of AIDS victims. Among the female students, 16 expressed concern about dying without children, leaving orphan children or infecting their babies. These concerns gave rise to contradictory expressions. Sarah, the unmarried doctor in story 3, is dying. She reflects: 'I do not mind dying – but dying without leaving a child on the earth, I am lost forever. God will have cheated me. God said "go out and multiply", why can't I be allowed to multiply?' This fictitious quotation is almost word for word the same as a real statement from an AIDS sufferer mentioned later in this chapter.

AIDS and abnormal times

That these writers saw the epidemic as unprecedented is witnessed by the use of the LuGanda term ***namuzisa*** – 'the one who causes extinction'. The epidemic is experienced as a phenomenon that defies common sense and the expected rules of nature: it is an event for which the past is not an adequate guide to the present or the future. The apocalyptic tone is caught by the proverb used by some people in Buganda to describe such extraordinary times: 'While some people are wandering about, the cat gives birth to a dog, the sheep climbs on to the roof of the house'. In a society like Buganda in which most people have been exposed to Christianity, we should not be surprised to hear such echoes of the Book of Revelations.

The apocalyptic tone that entered into some of the essays is continued in the following two comments:

> In ten years' time, seventy per cent of the people in Rakai will have died. There will be no one left to work. After a man who has ten acres sees that his sons are dying, what is there left to do but to leave the land fallow?

> In coming years, the country will be dominated by animals and trees . . . in future there will not be a healthy living person in Rakai.

These are comments which, in view of the evidence to be presented throughout the book, may be exaggerated but are also perceptive. In all the essays it is clear that these adolescents are well aware of how to reduce risk by changing behaviour. The various messages of public health-education programmes of such organizations as the AIDS Control Programme, the African Medical Research Foundation (AMREF), the Red Cross and the churches therefore do appear to be getting through to them. However, knowledge is one thing, a change in sexual practice is quite another.

Expectations and crisis in Buganda

In the previous section the perceptions and explanations of AIDS in Buganda society were examined. Here, the ways in which AIDS is perceived to affect people's expectations of life are discussed. In any culture people can be assumed to have a hierarchy of expectations. These may be threatened by a crisis that will force a reordering of expectations. The highest expectation might be self-respect and a sense of worth created by the giving and receiving of affection. Another, lower in the hierarchy, may be an acceptable standard of living now and in the future; while still lower ones may be a minimum food intake, basic shelter and short-term survival. Any crisis forces a person to reappraise these, perhaps to rearrange them and to abandon some at the expense of others.

Where AIDS has caused considerable mortality, expectations of the normal rewards of life have been put into jeopardy. Consider this quotation from a man living in the south-eastern Rakai District:

With AIDS we are now living on the front line just as we were during the Liberation War of 1979. Then we were prisoners of that war, now we are prisoners of AIDS. Those of us who are ill are under sentence of death.

The ways in which this all-enveloping crisis, from which there is no escape, threatens these expectations can be illustrated by the terms used to describe the disease such as ***mukenena***, the one that drains; ***lukonvuba***, an incurable disease; and ***mubbi***, the robber. These three terms for AIDS indicate how many of the expectations of normal life are threatened in a number of ways:

1 The threat of the disease robs adults of the expectation of sexual fulfilment and of marriage. As we have said, condoms are little known in these rural areas, and are usually unobtainable. Sex is therefore set about with extreme anxiety. The expectation of sexual fulfilment is either completely abandoned in favour of celibacy, or maintained by spurious explanations of why the disease will not be transmitted in particular or future imaginary sexual encounters.
2 AIDS puts a serious strain on any permanent or semi-permanent relationship. Fears of what someone may bring to the bed of their partner is a constant source of worry. Unwanted sex forced upon women is very much a live issue, discussed in women's informal groups.
3 If young people die childless, they are robbed as individuals as they leave no trace. For example, 'I don't mind dying, but to die without a child means that I will have perished without trace. God will have cheated me.'
4 In the same way, whole families feel themselves to have been robbed as they face extinction from the communal consciousness.
5 The disease deprives the elderly of the expectation of the care of their children in old age, and ultimately of a correct burial.
6 It robs children of their parents, of the love, training and security that they provide (see Chapter 7).
7 The disease drains not only the sufferers, but the resources of the family who look after them and who survive after they have died (see Chapter 6).

In these ways AIDS threatens almost every aspect of the normal expectations of family life from birth to death and beyond in the sense that proper burial and memory of the individual in the collective consciousness are threatened. The disease will also affect other expectations lower in the hierarchy, because economic security may be threatened. Widows and orphans may face insecurity of ownership or access to land and property. Food supplies may be disrupted because of a lack of labour and cash (see Chapters 6 and 8).

Conclusion

This chapter has suggested that individuals in Rakai feel they are living in abnormal times. They have had to struggle with developing explanations of those times. In the worst affected areas, original explanations through the idiom of witchcraft have rapidly been replaced by others that take account of medical facts or in which a syncretic combination of explanatory discourses coexists. In part this reflects the success of government educational efforts. However, in so far as such information is known, it does not necessarily result in behavioural changes. Instead, people (particularly some men) rationalize their way out of the full implications of the disease through a variety of other accounts of how the disease may be avoided, or by adopting an aggressive fatalism. In the case of adolescents, knowledge of the mode of disease transmission is widespread, but whether or not these young people will change their sexual behaviour remains to be seen. However, their essays seem to suggest they may be deflecting some of their fear and uncertainty into anger at those (mainly élite) people whom they perceive as having only themselves to blame and who are threatening the welfare of the country by wasting their privileged education.

There is a complex range of explanations and actions. People give different weight to different parts of the particular chain or chains of causal explanation which they choose to help them to cope. In part their choice of explanatory vocabulary and of the weighting they attach to different types of explanation appears to be linked to their position in society. Men choose escape hatches, women worry about their children, adolescents blame others and make moral judgements while focusing their concern partly upon the effects of wasted investment as they look towards a new, uncertain future. Almost everyone also recognizes that AIDS threatens all the major expectations of people's lives – from sexual fulfilment, to marriage and having children, to being cared for in one's old age by one's children, through to having a proper burial and being remembered in the community's consciousness after death.

Such rationalization and explanation is not unique to Uganda or to Africa. Shilts (1987) describes similar mental processes among homosexual men in San Francisco in the early 1980s. He reports the following types of response to the epidemic:

> There were the 'What Crisis?' types, who denied there was an epidemic at all, as opposed to the 'Nervous Nellies', who were paralysed with dread. The 'Ozzie and Harriets' had settled into monogamous relationships, while the 'Superman' types tricked on, convinced they were somehow immune to AIDS. The 'Doris Day' types invoked fatalism to rationalize their continued cruising, singing 'Que sera, sera' . . . [and] the last category, 'The Utterly Confused'. (Shilts, 1987, p. 377)

Notes

[1] This material was collected by Dr Christine Obbo. She would not necessarily agree with the analysis.

[2] The essays were written by pupils at Kakooma Secondary Schools No. 4 and 5.

Notes on further reading

Useful reading on the issues discussed in this chapter, not restricted by any means to Uganda, includes: Shilts, 1987; Gluckman, 1955; and above all Douglas and Wildavsky, 1982.

4

Coping with AIDS: the Downstream Social and Economic Effects

This chapter first analyses the AIDS pandemic in Uganda in terms of the concept of disaster, drawing some comparisons with other crises to explore to what extent theories of disasters and coping mechanisms can throw light on the impact of the disease. A discussion then follows on the ways in which people cope with the illness when it strikes them and their close relatives. How does the disease confront people's expectations and hopes? If there are conflicts between those of different people, whose are sacrificed and whose survive? Lastly, a framework is presented for analysing the economic and social structuring of vulnerability to disaster impact and access to resources for coping with it.

AIDS as a disaster

As Chapter 1 has briefly discussed, AIDS differs in some crucial respects from other types of disaster. We distinguish it as a *long wave* disaster because the AIDS pandemic does not take the form of a discrete event with recognizable stages and responses. In Chapter 3, it was suggested that when people are confronted by the uncertainty associated with a crisis they will undertake a range of 'experiments' as their experience of the new situation increases. Such experiments can be seen as attempts to reverse the balance between the known and the unknown in a search for normalization. In sudden disasters, such as earthquakes, where established coping mechanisms are inadequate, this period of experimentation may be brief. In the case of a disease such as AIDS, where the nature and epidemiology of the disease means that the onset of 'abnormality' is very gradual, but where the rapidity of spread increases as a critical mass of ill people is reached, the period of transition and experimentation may be quite long and delayed. The duration of the crisis, whether long or short, will provide advantages and disadvantages from the perspective of developing coping mechanisms whether these are local mechanisms (for example, new ways of caring for orphans) or society-wide policies (for example, allocating resources to a health budget that takes into account large numbers of AIDS sufferers).

This is so for three main reasons. Firstly, *a priori* strategies for preventing the disease require an accurate medical explanation and effective changes of sexual practice. For a variety of reasons, the perceptions, explanations and changes in practice are not in place, as Chapter 3 explained. Secondly, coping with the main immediate downstream effects (illness and death) involves well-established coping mechanisms acted upon in relation to each individual death. It is only when the gross scale of deaths gradually becomes apparent that coping at the margin is seen to be inadequate. In other words, the scale is unprecedented. But this is not necessarily perceived by individuals, for these inadequacies occur at the level of the whole community or region, rather than of the household or family where one or even two deaths may occur in quick succession in 'normal' times. There is now a widespread realization that the pandemic is unprecedented in most parts of Uganda and especially in Rakai and Masaka Districts.

Thirdly, AIDS can be described as a *long wave* disaster because it is a disaster that is a long time in the making and in which the major effects have already begun to occur long before the magnitude of the crisis is recognized and any response is possible. Parallels may be drawn here with similar long wave disasters such as global warming or the effects of acid rain. In such cases, the disaster does not appear to the affected society as a discrete occurrence with recognizable trigger events, which can be used to mobilize action, such as would be the case with an earthquake, a volcanic eruption or a devastating flood. Even a drought and the subsequent famine it may trigger both have recognizable onsets, although, as is well known, famines may take a relatively long period before their full effects are apparent. The broad differences between these different types of disaster are represented schematically in Figure 4.1.

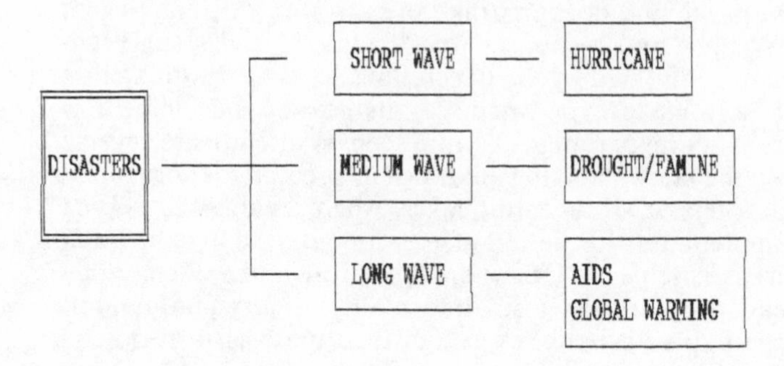

Figure 4.1 Short, medium and long wave disasters

Two rather different ideal types of coping responses to disaster can be distinguished, although in reality they overlap. The first involves *a priori* risk-aversion strategies. These provide ground rules for pre-emptive actions to minimize the probabilities that the disaster event will occur in the first place. Examples of such tactics include the production strategies of 'risk-averse' farmers and pastoralists. Thus, in such communities, multiple cropping, inter-cropping, multiple-enterprise livelihoods, grain storage and building up cattle numbers in well-watered years are ways of avoiding risks that are known to occur. There are also reproduction strategies in the face of high infant mortality, whereby a large number of births improves the chances of a minimum labour force and surviving children to look after parents in their old age. In all such cases, adverse conditions are expected and planned for.

The other type of coping mechanism is based on *ex post* strategies. These are the most common and well-noted and take many forms in different economies and cultures. The majority have evolved in response to risky physical environments, particularly to drought. They are often sequenced and identified with the different stages of a disaster as it unfolds: increasingly radical and desperate measures are brought into play as the situation worsens (Corbett, 1988). They also follow a sequence in terms of the level at which people are involved. Within the household, a reduction of consumption of staples, the sale of livestock or of items of wealth and the collection and consumption of famine foods are some examples. At a higher level, the seeking of wage employment, borrowing grain from relatives or mobilizing networks of obligations and rights from patrons and relatives may occur. Regional, national and international levels may also become involved in terms of public works and 'food-for-work' programmes.

The implications of AIDS as a long wave disaster are that the different stages of coping are extended and delayed. The first stage involves the recognition and explanation of the disease. Without the benefit of medical information being disseminated to the public by mass media or widespread public meetings, the only overt symptoms of a crisis may be an unusually widespread syndrome from which people are suffering and dying in large numbers. By the time these are recognized, seroprevalence is very high and the eventual scale of the disaster has enormously increased. Depending upon the rate of infection and the mean period between infection and the onset of recognizable symptoms of AIDS Related Complex, for every one person with these symptoms there will be many more who are already HIV+. Thus the raw material, the evidence, on which an explanation of the disaster is made, takes at least five years to appear. This may be said to indicate the length of the disaster wave, the period between the beginning of the disaster and the mobilization of coping responses.

Secondly, once people are recognized to be suffering and dying from the disease, the ways of coping are familiar: death and illness have established responses in all societies. Thus existing coping mechanisms are mobilized. It is only when an unprecedented number of deaths is suffered that existing mechanisms begin to seem inadequate. Again the adequacy of existing responses is driven by the spatial and temporal growth of the epidemic. As Chapter 2 indicated, in some parishes of Rakai District, over 2,000 parents have died in a total population of about 30,000 (Hunter, 1989a). The arrangements made for the care of the sick, support for close relatives, funerals, orphans and so on clearly cannot follow established patterns. In nearby parishes, where mortality is much lower, existing coping mechanisms obviously show fewer signs of stress. However, adaptations can only be made on the basis of experience, not by prior planning. Unlike some other disasters, such as famine or floods, *a priori* coping mechanisms and experimentation cannot easily be brought to bear. In the case of famine, the stock-piling of food and other short-term risk-averting strategies can be followed before the symptoms make themselves felt. The only provision that would parallel such a strategy in the case of AIDS would be a widespread change of sexual behaviour. In the case of such a long wave disaster, this change is only likely to take place when the disaster is already well-advanced.

Individuals coping with AIDS

Individuals can be categorized as being members of households that are of three types. First, there are *afflicted* households that have one or more members suffering from the disease, or who have already lost someone. Next there are *affected* households that do not have a member ill or dead from AIDS, but have received orphans from other members of the family or from neighbours. Then there are those households that are not directly touched by the disease.

In AIDS-afflicted households, where there is an ill member, people spend a lot of time in prayer at home, in church and at Catholic shrines. Some people, at great financial sacrifice, have gone in search of a miracle AIDS drug supposedly developed in Zaïre by a Zaïrian and an Egyptian doctor. Others have joined the bus-loads of people who have travelled long distances over rough terrain to obtain curative soil from a visionary called Nanyonga.

Household members who participate in these activities are either involved in nursing or they are visiting temporarily. In devout Christian and Muslim households the comforts of prayer and the clergy are actively sought. In other households the advice of the traditional diviner is preferred. In both types of household, people fear the spirits of those who died in anger or with a grudge. People who have died in anger are considered to be a threat to the health of surviving household members. However, households differed in the ways in which they coped with the emotions surrounding illness and death.

In devout households, if a patient was not already in hospital there would be eleventh-hour attempts to get them there. Panic was usually triggered by faint murmurs of the patient: 'Don't leave me in the dark' or 'Are you letting me die without medical care?' If the patient was left to die at home, the household might be haunted by misfortune. Thus hospital care was sought despite the heavy expense involved in returning the body from hospital to village for burial.

In less devout households, after pleas similar to the above, an elderly member of the household would volunteer to visit a diviner, acting as an intermediary between close relatives and the dying person. There were no local diviners in the research area, but there were some in adjoining counties in the district. Sometimes people preferred to go outside the district, believing that a distant diviner could give better advice than a local one.

Diviners were consulted to find out whether the patient was satisfied or had been slighted by and/or was angry with a household member. In cases of real or potential tension between the patient and care-givers, diviners gave advice on how it might be resolved. A hen, cock, goat or sheep of particular colours might be required and the purchase of herbal remedies necessary. Some diviners were paid immediately; others asked the client to come and offer thanks when the problem was solved. Thus the diviner plays an important role in the coping process; acting as a kind of family therapist and bereavement counsellor, perhaps facilitating a 'settlement' of the affairs of the household at the time of death. Such recourse to the counselling and curative skills of diviners and traditional healers is found in other parts of Africa (Thompson, 1990).

Care-givers listen carefully to the patients and many are diligent in meeting most of their physical demands. Whereas people are less unsure about the emotional state of anger, when a patient says: 'Get me medicine for the pain' or 'You have not told me the truth about this illness', they find it difficult to cope with a desperate fear of death, and beseechings such as: 'Don't abandon me.' People publicly admitted that because the AIDS illness takes so long to kill its victims, it exhausts both the patients and care-givers, in addition to reducing the patient to skin and bones. AIDS was therefore referred to as ***mukenena*** – the body shrinker. In this state the patient may be extremely demanding of the care-givers who are also overwhelmed with other family matters. This might be a source of unspoken tension as the following case reveals.

A patient who developed full-blown AIDS symptoms after giving birth died after two months. She slowly came to realize that she had AIDS but her educated family was having her tested for over half a dozen other illnesses – except AIDS. The day before she died she confronted her family about lying to her and to themselves. The physical and psychic drain necessitated consulting diviners at least for the survivors' peace of mind. The care-givers also seemed to seek reassurances about their negative feelings towards the patients. Sometimes the care-givers would express their frustration with the job but at other times neighbours and friends would articulate their precise feelings for them. For example, patients were said to be unfair if they 'appeared finished and yet hanging on to life'. People referred to 'having the will to live' as unfair to the living particularly the care-givers. But ultimately it is the incurable disease (***lukonvuba***) that is to blame.

However, for other AIDS sufferers who are not yet seriously dependent, tensions are not so pronounced and they continue as integrated members of the community; others withdraw from the daily life of the village at an early stage. The following cases illustrate such situations:

The Butcher
The local butcher woke up one morning with a rash on his neck and legs. Experience told him that this was a symptom of AIDS. He was so frightened that he became hysterical. He told all of his customers. Most of them stayed for a few moments to express sympathy. Some muttered 'this disease will wipe us all out'.

The butcher continued to work until he was too weak to carry on. People continued to patronize him as he was the only butcher in the village. Some came expressly to give him moral support. He spoke openly about his condition. When he finally became bed-ridden he insisted that his bed be put in the sitting room of his house so that he could receive visitors. Nine months after the first manifestation of his symptoms, he was still moving around the village, although it was an effort for him to do so as he became weaker.

The Trader
An itinerant trader, on finding that he had contracted AIDS, became a recluse. Although he continued to work, he stopped socializing with others in the community. Previously he had been considered a good conversationalist. People knew that for the last three years this man had been having a relationship with the widow of a man who had died of AIDS.

Coping with death

Next we turn to a discussion of the way people cope with the death of members of the family. We shall look first at how women cope with the death of their husbands. Widows may stay in their homes, migrate to urban areas, remarry or return to their parents' home. It seems that the majority of widows remain in their marital homes to support themselves and their children. They experience a decline in their standard of living because women in general have less entitlement than men to other people's labour. Brothers-in-law, even when they live nearby, offer minimal help. Furthermore, widows of migrants who had never legally acquired occupation rights in land, and widows living on land acquired by their husbands under occupancy rights but where the ownership has changed, have even less security of tenure and are sometimes threatened with eviction. In such cases, the landlords claim that the 1927 Busulu and Nvujjo law was a contract between the landlord and the husband, specifically *excluding* the wife and children (Mukwaya, 1954). Since the 1975 Land Decree during the Amin period, there has been considerable confusion and uncertainty on this matter (Government of Uganda, 1975). In the community from which this case material was gathered, to date two widows of migrant squatters together with their children have been evicted on these grounds. Widows who do not wish to undertake the heavy agricultural work alone and to continue to live in rural poverty migrate to the urban areas to improve their material well-being. They join other urban migrants in self-employment such as craft production and food or beer vending.

There has been speculation among élite women in Kampala that Rakai widows will be pushed by desperation into polygynous marriages and prostitution. This has not happened so far, but it may in the future as a result of fierce competition between widows and other single women for sources of income. Young widows aged between 20 and 40 are able to remarry single and unmarried, divorced or widowed men.

There were ten widow remarriages in the study community. Six involved widows who had migrated or returned to the area, four were of local widows. Widows who are young and attractive are assumed by many to be AIDS-free because many people in Uganda find it difficult to believe that somebody can be a carrier of the disease while remaining perfectly healthy (see Chapter 3). This can give rise to anxiety when marriage is being considered. For example, in five instances during fieldwork, prospective husbands were warned by concerned friends about the dangers posed to them by an AIDS widow. It appears that some of the men knew the risk and were willing to take it. Such men answered those whom they perceived to be interfering: 'What benefit do people get if they die late of old age?'; 'Bodies are all buried the same way; will they sew me into timber?' (***ndibajjibwamu mbawo***); 'Will I grow into mushrooms (and be edible)?' (***ndimera butiiko***); 'Who is never going to die?' (***ani ataliffa***)

These somewhat gnomic utterances appear to indicate fatalism in the face of a disease that people recognize to be fatal. However, not only fatalism is indicated. It must also be recognized that such statements indicate efforts to continue a 'normal' existence in circumstances that are rapidly becoming decidedly 'abnormal'. Such attitudes may appear perverse, but they must be seen for what they are: one way of coping with the presence of death in close association with sexuality.

Professional men volunteered in conversation that they regarded celibacy as a temporary solution until a cure for AIDS was found. By and large, marriage was seen as a buffer against AIDS. Religious messages in sermons and posters promoted 'love faithfully' as a precaution against AIDS. Both men and women enter into sexual relationships with, and even marry, someone who looks healthy on the basis that they must be AIDS-free. However, as we noted in Chapter 2, current seroprevalence among men and women between the ages of 19 and 35 in Rakai District makes marriage a very risky choice. These dangers are becoming more apparent to the people themselves. In the study community, there had been three cases in which women, after six months or a year of marriage, woke up in shock to find their husbands covered with the rash of ***Herpes Zosta*** and showing the usual symptoms of AIDS (persistent dry cough, weight loss, fever and diarrhoea). In all three cases the reaction of the women was to leave the home: their husbands had to depend upon other people for support and care. Young widows with two or three young children tended to return to their natal home. Fear of the exclusive burden of farm work, insufficient money to hire labourers and the desire to receive unconditional help with the children were given as reasons for this move.

Other widows opted for celibacy. They felt it not good to tempt fate since they had been lucky enough to escape AIDS with their deceased partner. In this predominantly Catholic community, most people affected by AIDS either as patients or as survivors of an AIDS-infected spouse have turned to religion for comfort. Widows perceive the nuns as their model of piety, celibacy and work. However, this is not solely a response to the disease, as celibacy has always been one of the strategies which Ganda widows have used to avoid involvement and problems arising from men who might be after their property.

We turn now to the way in which men cope with the illness and death of their wives. From our own fieldwork as well as from Hunter's major study of orphans (Hunter, 1989a and b; Dunn et al., 1991), it seems that many more men have died than women and that widowers are much rarer than widows. Men whose wives are ill and eventually die cope in similar ways, although access to economic support in terms of labour to work the farm and cash to enable the sufferer to receive treatment from diviners, healers and hospital is usually much easier than for women. It is probably true to say that they spend less time attending to the needs of their wives while ill (cooking special meals, staying with them in times of particular need). Care of the children can sometimes be arranged through the husband's sister and other of his female relatives. Refuge in the company of other men in bars is another way of sharing the burden.

Access to resources

So far in this chapter we have discussed the perceptions and explanations of AIDS in Buganda and how these differ depending upon the experience of the levels of morbidity and mortality in any one area. Within each area, men and women have had substantially different experience of the impact of the disease and their expectations of life in the future have been affected differently. In what follows, the substance of what people think about AIDS is linked to broader processes in the economy and society.

Every society experiences social and economic change and a disaster such as AIDS impacts upon these ongoing processes. An heuristic model is presented in Figure 4.2 which formalizes the impact of AIDS on the changing patterns of access to resources involved in earning a living.

It shows how individuals (usually as a part of households and more extended familial networks) gain access to physical resources (e.g. land, labour, agricultural implements, shelter, etc.) in order to combine them to support themselves. It can be understood as a more general expression of the model developed by Sen (1981) which uses the concept of

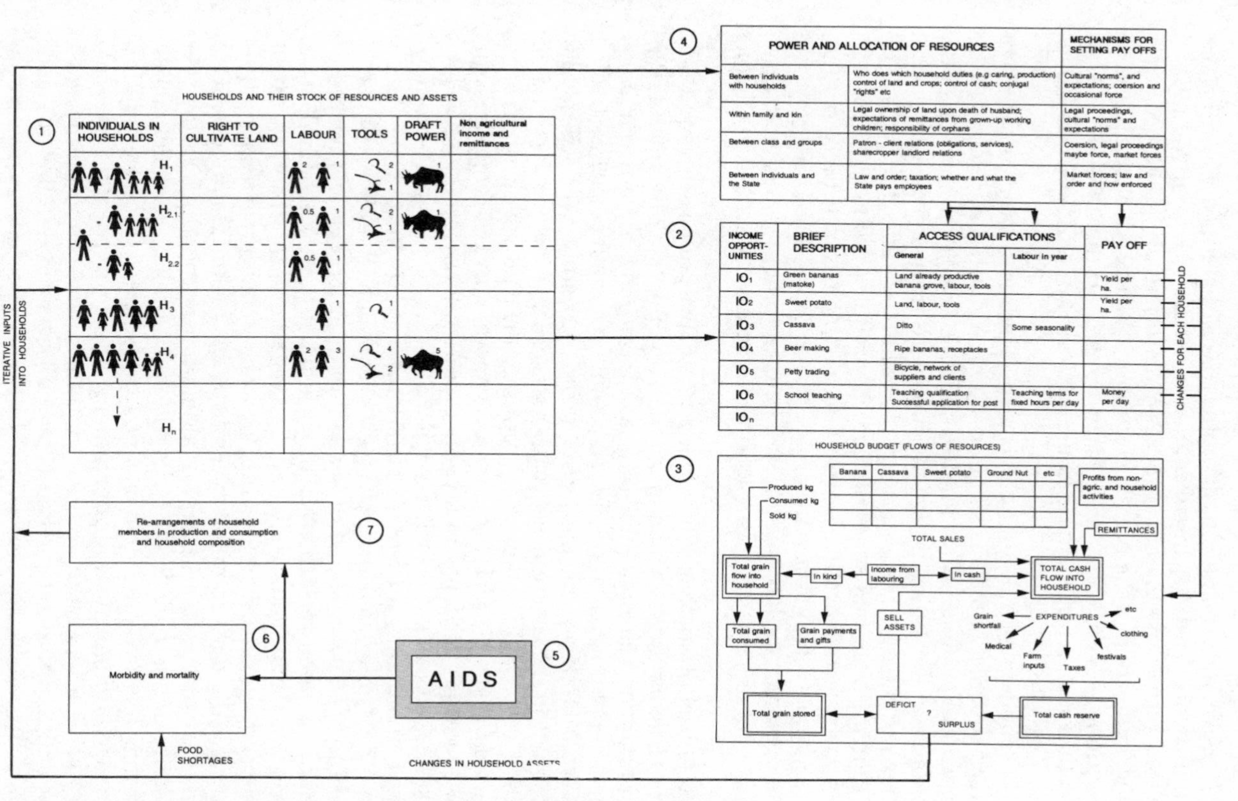

Figure 4.2 Access to resources and vulnerability in the face of AIDS.

entitlements in an explanation of famines. However, this model indicates vulnerability to the downstream impact of AIDS on an individual and household basis. This model of access was originally developed for a computer simulation of the Nepalese economy (Blaikie, Cameron, Fleming and Seddon, 1977) and has since been adapted to the analysis of land degradation and its impact upon households in different rural economies (Blaikie, 1985).

In this model, each household is conceived as having an array of resources, broadly defined as those economic, financial and social assets which, when combined, allow for the production and reproduction of the household and the satisfaction of the culturally defined needs of its members. Box 1 in Figure 4.2 represents an array of households, each with different resources. Note that some of these households are polygynous and that others are female-headed, representing a point in time between partnerships (serial polygamy) when there is no current male to provide any substantial resources to that household. There are other types of household which are not represented in Box 1. Some of these are described in Chapter 6.

Each household is conceived as viewing an array of possible economic activities listed in Box 2, which in combination provide a livelihood. Typically in Buganda, most of these income opportunities are different crops, but there are also important non-agricultural opportunities such as trading, agricultural labouring on the farms of others, selling beer, fishing and employment in the government sector. Each of these income opportunities has an access qualification. For example, the cultivation of bananas requires access to a previously planted banana grove and the necessary labour to prune the plants, mulch the ground underneath and weed it if necessary. Annual crops will also have a typical seasonal profile of labour demand and will require seed, land and labour.

It can be appreciated that when an event such as AIDS reduces the availability of labour, one of the most important access qualifications is the ability to mobilize sufficient labour at times of peak labour demand. Other income opportunities have lesser access qualifications and are usually over-subscribed and therefore are less well remunerated (agricultural labouring is the most common example in this region). Trading activities will have a variety of necessary qualifications, depending on the scale and nature of the trade. Some types will demand that the trader is male; others may require considerable capital or the ownership of a bicycle or even a lorry. Employment in government service will require a particular level of schooling, and so on. Each of these income opportunities has a pay-off and contributes to the household budget (stylized in Box 2). Gender is often an important access qualification. Indeed, many income opportunities are denied to women. The economic insecurity of women is a major factor, not only of the way in which AIDS impacts on Ganda society, but also in the pattern of sexual practice and the spread of the disease in the first place. For example, marriage or the possession of a male partner can be viewed partly as a means of livelihood for some women in a situation where other more conventional economic activities are denied them. This issue is discussed further in Chapter 5.

Pay-offs for each income opportunity are determined by a number of variables. Yields of the crops grown, the price the producer is able to secure for produce in the market and margins in trade are the most common. However, there are also other important non-technical and non-economic factors that decide the level of pay-off, especially where it is subject to prevailing power structures. For example, a husband may constrain the extent to which a woman enters the market to sell the goods she has produced. As will be explained in Chapter 5, a woman was conventionally given a banana grove to cultivate, but she may have to sell the bananas through her husband and only retain part of the proceeds for herself. The degree to which elderly relatives may receive remittances, labouring help on the farm or other assistance, is subject to expectations and norms that lie outside the market.

These factors collectively have been labelled 'power and the allocation of resources' and are represented in Box 4. They operate at a number of different levels. Within the household, they depend principally on gender, seniority and relationship with those with access to income opportunities. Within the wider family and kin networks, but outside the household, these typically involve obligations of material assistance on a regular basis and at life ceremonies, the temporary care of relatives' children and the settlement of property issues at times of death, divorce or other change in life status. Some important processes involved in the allocation of resources also involve the state. The level of the maintenance of law and order is one example. As will be seen in Chapter 5, the upheavals in the mid-1970s involved the collapse of the existing distributive and trading network, resulting in a massive upsurge of smuggling, rapid inflation and violence. The way in which land law is interpreted and upheld, particularly in disputes between widows and the deceased husbands' relatives or between cultivators and landlords, all have profound effects upon both access qualifications and pay-offs.

The sum of these pay-offs together constitute a livelihood, and flow into the household budget, as shown in Box 3. Thus the household budget consists of inflows in the form of crops and materials for use and sale, with or without further processing. Typical examples of the latter are beer-making, the weaving of mats and containers, and bark processing to produce a fabric used for clothing and funeral shrouds. Cash may also enter the household from the sale of crops (in Buganda these are predominantly coffee, bananas and sometimes Irish potatoes). Outflows consist of consumption of foodstuffs, payment of wages to labourers, fees to the Parent Teachers Association (which can be particularly onerous, see Chapters 1 and 7) and the money spent on other purchased household items.

The stock and flow situation of the household budget can be monitored at certain time intervals, most usefully at the beginning of each agricultural season, when stocks are typically at their lowest. If the net situation is in surplus, then the household may accumulate wealth, either temporarily, or may invest in further income-earning opportunities. In Buganda at the present time trade is the most common investment, since investment in coffee production gives little return. The state has a legal monopoly of the purchase of all coffee. The price paid to farmers is most unattractive and payment is often delayed for many months. Other productive opportunities have mostly declined during the past 20 years of unrest. If the net situation is negative, the household must reduce consumption, and/or borrow or somehow secure enough of their basic needs to survive.

AIDS thus directly affects the numbers of people in a household, whose structures may change with desertion of a sick partner or the sending of children to relatives. With the onset of the more debilitating symptoms of the disease, the sufferer can no longer work productively and the access qualifications for producing some crops may become too high. Remittances from non-agricultural income opportunities may cease upon the death of a son in government employment or in trade. The household budget will suffer a sharp decline in cash and possibly food crop inflows. Children may have to be taken out of school, and other purchased household items will simply become too expensive. These multiple impacts are discussed in Chapter 6 and the special case of orphans in Chapter 7. Finally, in the longer term, the disease may begin to affect the entire system of production as labour becomes scarce. This major issue of the impact of the disease on farming systems in discussed in Chapter 8.

The model we have presented shows the ways in which households confront decision, are formed, grow, split up and disappear. New members are added through marriage and birth; others are lost through marriage, divorce, death and out-migration. Their economic fortunes may improve or decline through time. It is an iterative model, illustrating the way in which some households accumulate and others disinvest. Into these ongoing processes of earning a living and of longer term agrarian change, AIDS is introduced. What this model illustrates is the general processes affecting household and individual survival choices. Thus it is not a model that begins from conventional assumptions about the household as a decision-making unit, treating it as though it were internally unified and as though the environment in which decisions are made were decision-neutral. Instead, it attempts to capture the nature of power and of how the allocation of resources takes place, but it cannot start to analyse the wider structure of social, economic and political relations in which everyone is imbricated. These sets of relationships beyond the individual and the household make them more or less vulnerable to the impact of a crisis. It is for this reason that Chapter 5 moves beyond the proximate reasons for vulnerability to the impact of AIDS and provides a broader account of the circumstances of Uganda. It also provides a context necessary for understanding the circumstances of risk and vulnerability within which individual and household choices are made. Put simply, it is not just that individuals and households make decisions and take risks or are more or less vulnerable. Rather, it is that some environments are more risky than others, not only for decision-making about the production and reproduction of the individual and the household, but also in the particular downstream effects likely to develop as a result of an epidemic disease.

5

Vulnerability and the Ecology of Risk in Buganda

Risk behaviour and risk environments

Civil disruption, war, smuggling, unequal access to economic resources: all these factors have contributed to the particular form the AIDS epidemic has taken in Buganda. By the early 1980s, the social and economic landscape of Uganda – and of Buganda in particular – exhibited symptoms of differentiation and contradiction that provided a fertile environment for the spread of HIV. This chapter provides an outline account of the way in which that environment came into existence and of how within the social and economic relations that characterize it, the question has become less one of certain sexual behaviours being risky, but of all sexual behaviour being risky because the environment itself is one of high risk.

Much has been written in the AIDS literature concerning 'risk behaviours'. The WHO classification of global epidemiological patterns summarized in Chapter 1, is conceived in relation to types of *behaviour*. Three patterns were distinguished. Such a classification provides a brief descriptive typology of the main modes of transmission. It is useful as a summary epidemiological description, but it accords primacy to the act of transmission through heterosexual or homosexual intercourse, blood transfusion or intravenous drug use. The classification is not designed to locate the modes of transmission within a specific social context and relate them to processes that may mean that a specific act may be high risk in one social, political and cultural environment but not in another.

For example, the spread of the virus through sexual tourism in parts of south-east Asia, in the brothels of Bangkok and Manila, reflects the relative incomes of prostitutes and their clients. The gross income inequalities between prostitutes and their often foreign clients indicates on the one hand the high disposable income available to men in the countries of North America, Western Europe and Australia, and the rural and urban poverty in Thailand and the Philippines that drives women into prostitution. Or, to take another case, the spread of the virus among homosexuals in the USA in the 1970s took place among a relatively high income group involved in a political reaction against a discriminatory moral milieu. High incomes and protest led to the promiscuity associated with the 'bath houses' of San Francisco in the 1970s (Shilts, 1987). In each case, it was individual behaviour in a particular social, economic and cultural context that shaped the epidemic. To focus on individual behaviour alone, as does the WHO three-pattern schema of the pandemic, without giving due weight to the conditions within which that behaviour occurs, is to tell only a part of the story. For the purposes of simple exposition of the spread of the pandemic, such a basis of explanation is satisfactory; but it can also provide a clear target for scapegoating, since it identifies those who are most vulnerable to infection and therefore who are 'dangerous' to the rest of society. Simply, there is a social and epidemiological context to transmission. To give an example unassociated with the disease: vehicle theft is a common occurrence. Is it that the owner has indulged in risky behaviour by owning a car in an inner-city area in the US or Europe, or is it more informative to suggest that owning a car in that particular *place* is the risk? Insurance companies take the latter view, setting their premium rates in the light of the observed spatial distribution of risk.

Thus, in certain areas of Uganda, *any* form of sexual intercourse amounts to risky behaviour, but in any other place it may not be high-risk behaviour at all. We have seen in Chapter 2 that in Rakai District of Uganda, men and women aged between 20 and 30 are around 40 per cent seropositive. Such an observation does not tell us a great deal about why heterosexual intercourse has taken on this complexion there. More useful is an understanding of the social, economic and historical factors that have produced this particular spatial distribution of risk. In other words, it may be useful to balance the concept of risky behaviour with that of a risky environment or area and to recognize that particular social, economic and cultural characteristics of a society may produce environments in which risk is high and which can thus be said to be ***risk foci***. Buganda is one such focus. The reasons are explored in the next sections.

The recent history of Uganda: divisions and contradictions

The story of Buganda has to be seen against the background of Ugandan history and society as a whole, as a British colony and as a politically independent nation. The early part of that history revolves around the cultivation of two major cash crops – cotton and coffee – in the south of the country and the reorganization of pre-colonial society to facilitate their production. The key to this was Buganda.

The opening of the Uganda railway in 1902 linked Mombasa to Kampala and reduced the cost of transporting commodities produced in the African interior. When British forces arrived in Buganda in the 1880s and 1890s.

they found a centralized, 'feudal' state with an effective ruling class. In extending their rule, the British co-opted this ruling class by means of a land tenure reform introduced in 1900. This converted feudal tenure (with its checks and balances of reciprocal obligations) into individual tenure. Thus all the powerholders of Buganda, from the Kabaka (king) down to the lowliest parish chief, were allocated parcels of land. The land was measured in miles – resulting in what became known as the ***mailo*** system of land tenure. Thus was Buganda society converted from a feudal system into a society of landlords and tenants. By means of rent (***busulu***) and tribute paid in cash crops (***nvujjo***), the landlords were able to establish their economic, political and social domination. Cotton production was at the base of this system.

By the late 1890s, the Lancashire cotton industry had lost its assured sources of supply in North America and Egypt and was facing competition from factories in Germany, India and Japan. The British Cotton Growers' Association was formed in Manchester in 1902, its goal to encourage the cultivation of cotton throughout the Empire to ensure raw cotton supplies for the Lancashire industry (Barnett, 1977). Cotton seeds were first brought to Uganda in 1903 and its cultivation spread rapidly throughout Buganda. By 1928, the Busulu and Nvujjo Law was passed. This limited the exactions landlords could obtain from their tenants, effectively neutralizing their power and guaranteeing tenant occupancy as long as they grew cash crops. The changes resulted in considerable rural prosperity. However, the British administration's aim was to keep Africans in the rural areas, producing cotton. They were discouraged from entering into trade. Instead trade (and ultimately cotton-ginning) was put into the hands of migrant 'Asians' who were seen as both a link between the British and the people of Uganda and also a barrier to the development of an indigenous Ugandan trading class. In addition, as a 'foreign' stratum, it was assumed that the main interests of the 'Asians' would always remain in import and export rather than in the internal development of the country. Thus by 1938, 'Asians' came to control cotton-ginning, wholesale trade and also commerce (Mamdani, 1975, pp. 32–3.). Africans remained in the agricultural sector, but in one that was undergoing rapid change.

The changes in land tenure consequent upon the 1928 Busulu and Nvujjo Law permitted the development of a class of wealthy peasants, now selling directly to the 'Asian'-owned ginneries without having to meet obligations to the traditional feudal rulers. These people cultivated through their own efforts and through employing hired labour but were unable to become truly capitalist farmers as the land laws did not permit the sale and purchase of land. Elements of a usufructory system remained, limiting their development as capitalists. But with the introduction of coffee in the 1920s and 1930s, the incomes of these rural entrepreneurs were further swollen. The interests of such an economic class conflicted with those of others: the owners of the ginneries, the 'Asian' traders and merchants, the traditional artistocracy in the person of the Kabaka and his relatives and with their labourers – local, Rwandan and Burundaise. By the late 1940s, this had become organized political opposition and had taken a violent form in riots against the property of the government of Buganda and the 'Asian' owners of the ginneries and commercial premises. This was soon followed by political organization among the small African commercial class who saw themselves blocked by the predominance of 'Asians' in this sphere. The 1950s in Buganda saw a succession of violent incidents and boycotts aimed at 'Asians'; by the 1960s, the language of political protest had become a racial one – Africans against 'Asians'. This was to have dramatic results in the next decade.

Prior to the colonial period, Buganda was one of several interlacustrine societies in which feudal kingdoms had developed. During the colonial period it retained a degree of autonomy within the structure of the British Protectorate and retains a clear identity to this day. In the move to independence there were pressures from within Buganda for secession from Uganda. The years after independence in 1962 were characterized by continuing struggles between the central government in Entebbe/Kampala and Buganda, leading by 1966 to the ending of its remaining autonomy by the government of Milton Obote.

Thus we see in Uganda as a whole by the late 1960s the emergence of certain clear lines of division – between Africans and 'Asians' and between Buganda and Uganda. This situation was to be exacerbated by economic circumstances on the wider world stage. Between 1962 and 1969 the international prices of both cotton and coffee, Uganda's main exports, fell dramatically. In these circumstances: 'With prices soaring high and political discontent spreading among all classes, the governing bureaucracy was compelled to act decisively . . . [by a] . . . move to the Left' (Mamdani, 1975, pp. 48–9). A rash of banking laws and 'nationalizations' was enacted, aimed at increasing the power of the state personnel, a power that came to be exercised through patronage and pay-offs. This strengthening of the power of state officials was not in the interests of the small class of African traders and merchants who saw themselves in competition with the 'Asian' petty traders. Under the 1969 Trade Licensing Act, the government made a distinction between 'Asian' petty traders (who were encouraged to leave Uganda) and members of the larger 'Asian' commercial class who were granted citizenship and thus encouraged to stay. This made it clear that: 'the governing bureaucracy preferred an Asian to an African petty bourgeoisie. The latter was a political threat, the former was not' (Mamdani, 1975, pp. 52–3). So yet another contradiction had appeared in Uganda's political and economic life: that between the governing class and the African petty bourgeoisie. The latter looked to the army for support and the first attempt at a military coup occurred in 1966. Milton Obote's government attempted to neutralize the army by isolating individuals seen as a key threat. One of these was a senior officer called Idi Amin. Attempts at neutralization through military reorganization were ineffective and Amin finally effected a coup against Milton Obote in 1971. The main supporters for his coup were the members of the African petty bourgeoisie who saw it as a way of defeating the state personnel and the 'Asians', both groups which frustrated their interest in moving from being petty bourgeoisie to grand bourgeoisie. An additional source of support for Amin was from the north of the country (his area of origin) which had always seen itself as being relatively dispossessed compared with the more prosperous south - another source of division and friction in Ugandan society.

The economic mismanagment and terror of the Amin period are well known. Their importance for understanding the creation of a high-risk area in the sense that we are using it, lies in the creation of the ***magendo*** (literally the 'pilgrimage of greed') economy. In addition to the expulsion of the 'Asians' and thus the destruction of established systems of distribution, Amin's economic policies froze agricultural prices and drew large quantities of cash crops into the black market. Here lay the roots of ***magendo***. This was of particular importance in Buganda which supplied much of the food needs of the major towns of Kampala and Jinja. In addition, the ***mafutamingi*** (literally 'fat ones' - the operators in the illicit economy) also began to move into transport. The essence of the ***magendo*** economy was the smuggling of coffee, paraffin, sugar and gold out of the country, vehicle spares and other necessities into the country and food within the country. The profits from this trade were invested in housing, commercial buildings and land (Rothschild and Harbeson, 1981; Bond and Vincent, 1990). The atmosphere of these times has been graphically described as follows:

> The basic supply-route from the port of Mombasa, through Nairobi to Uganda, ran like a great artery of corruption from western Kenya to Kampala, on north to the Sudan, on west and south to Rwanda, Burundi and Zaire. Long stretches have been beaten to pieces as hundreds of trailer-trucks pound continuously up and down the roads. The tough drivers and crews, who are paid overtime and danger money, changed their Kenyan shillings at the border. They are often delayed for days, drinking in the bars, eating in the hotels . . . sleeping in the brothels which line the route. From this main artery corrupting tentacles of the black market with its illicit deals and violent transactions penetrate into the Uganda countryside, pulling into its stream the desperate, the opportunistic, and the down-and-out (Southall, 1980, pp. 627–56).

And, as Buganda, and Rakai District in particular, fronts the lakeside and had many casting-off points for the journey to Tanzania, it was inevitable that much of this 'corruption' would come to rest in that part of Uganda.

However, this was not all. When Amin was overthrown in 1979 by the Tanzanian army, the fighting and the invading army passed through parts of Buganda, bringing Milton Obote back to power at the head of what was essentially a military government. During the next period of Uganda's history, known in Uganda as 'Obote II', the decay of the state structure continued and, some say, the oppression and violence of this new military government was even worse than under Amin's rule. It was certainly a regime that allowed little in the way of 'normality' to reappear in people's lives; the illicit economy continued to thrive as a vital part of people's survival strategies. Finally, in 1986, Yoweri Museveni, at the head of an armed movement, the National Resistance Army, defeated and expelled Milton Obote and took office as president of Uganda.

Since 1986, President Museveni's government has undoubtedly established a higher level of civil and political order than was the case during the preceding two decades. However, Uganda remains a troubled society. There have been signs that both old and some new divisions are reappearing. In the east of the country, the Uganda People's Army has been active around Soroti. There have been ambushes of government soldiers and attacks on the civilian population. In May 1991, the Ministry of Relief and Rehabilitation reported that about 160,000 people in this area were living in camps for the displaced. In Gulu and Kitgum in the north of Uganda, the emergence of the Uganda Democratic Christian Army in mid-1990 led to increased disorder and movements of people. The 1989 crop failure in Karamoja created serious local food shortages, resulting in heavily armed Karamojong raiding parties infiltrating Acholi, Lango and Teso during 1990. Also in the north, the effects of the war and disorder in Southern Sudan have been felt: Toposa raiding parties have pushed across the border into Uganda, displacing several thousand Ik people within Uganda. On the western border with Rwanda, the Rwenzururu Movement and the National Army for the Liberation of Uganda have been active. This and the response by the government's National Resistance Army have created instability, population movement and potential food insecurity. Such political and military events contribute to the conditions for continuing and further population movement and insecurity.

From this brief survey, we see that Ugandan society has had, and continues to have within it, certain social, political and economic cleavages – Buganda and non-Buganda, north and south, African and 'Asian', large rural farmers and agricultural labourers. As these divisions became more pronounced in the period from 1960 to 1986, so they constituted structures of vulnerability within the social body: structures that could facilitate the spread of epidemic illness. There was, however, one additional division in Ugandan society which is of consequence but which has so far not been mentioned: that between men and women. As will be seen in the next section that describes the detail of the development of Rakai as a focus of risk within Buganda and Uganda, this is of the greatest importance.

Rakai District – the development of a risk focus

It has already been noted that, according to local people, AIDS is supposed to have made its way into the lakeside communities in Rakai from Tanzania during the late 1970s and early 1980s. It is certainly true that the area of Tanzania immediately across the border, in Bukoba and Kagera, also has a significant AIDS problem (Rajab, 1991). It appears that a certain set of conditions may have contributed to the rapid spread of the disease in this area. The immediate causes of this rapid spread are directly related to the pattern of sexual contact. These can be established only very approximately through epidemiological modelling since empirical studies on sexual networks in Africa are even more inadequate than those in Europe and the United States. Doubt remains as to the relative importance of different parameters, but two are very important. These are the mean rate of sex-partner change and the variance of sex-partner change between sexual-activity classes (see Anderson, May, Boily, Garnett and Rowley, 1991). The movement patterns of sexual partners for reasons such as journey to work and seasonal migration also constitute another important set of variables determining the spatial diffusion of the pandemic. There are other important variables, too, such as the age of sexual partners, which influence the demographic impact of the disease.

Here we outline some of the social and economic circumstances behind the epidemiological pattern of the disease in Rakai. These circumstances include: the high mean rate of sexual-partner change, in some circumstances a high mixing of partners across sexual activity classes and a high spatial mixing of sexual partners. It is difficult to provide a rigorous account of the impact of social disruption over the last 20 years upon the patterns of sexual activity most associated with the rate of spread of AIDS. Unfortunately, there is virtually no information about the patterns of other sexually transmitted diseases (STDs), so they cannot be used as 'markers' to trace the impact of rapid, disruptive social change upon sexual practice in the past two decades. Therefore, in what follows we propose a set of reasoned hypotheses. These revolve around four related socioeconomic and historical features of Uganda as they affected Rakai: the existence of ***magendo***, warfare, general civil unrest and the status of women in Buganda.

With Amin's expulsion of the Asian community from Uganda, the formal exchange and distribution arrangements for cash crops and other commodities broke down. By 1975 the Ugandan formal economy had collapsed and had been replaced by the underground economy. Access to industrial consumer goods was through smuggling agricultural produce to neighbouring countries, the return with the merchandise and charging exorbitant prices, which some people were willing to pay. Smuggling was recognized as a hazardous occupation which added to the costs of trade in both money and life. Smuggling and black-market currency exchanges took place at all points along Uganda's international boundaries – West Nile, Kigezi, Bugisu and Tororo. All along Lake Victoria, fishing villages were transformed at night into smuggling ports, visited by motor launches and canoes. Coffee was smuggled to Kenya and Rwanda on long-distance lorries. But business was best across Lake Victoria into Kenya (Green, 1981; Mamdani, 1988). Coffee smuggling was dangerous but it was the quickest and best way to be paid in foreign currency. President Amin, who, it was rumoured, was himself airlifting coffee to America and Britain with the help of mercenaries, issued several decrees aimed at reducing coffee smuggling. An anti-smuggling security force operated by day and by night on Lake Victoria and coffee smugglers were shot on sight. This shifted smuggling to the trucks that operated along the Mombasa-Nairobi-Kampala-Bujumbura highway, and also to the hitherto remote fishing villages on Lake Victoria such as Kasensero in Rakai District. Recalling those times in 1989 and 1990, one respondent said: 'People had money and could exchange it into any currency. Young men dropped out of school to earn the easy money. Women joined in because it was much safer for them than for men.' These activities created a demand for the provision of food, lodging and sex at truck-stop townships, border towns and the smuggling villages on Lake Victoria into which lake traffic carried smuggled goods (see also an analysis of the role of trading patterns in the spread of AIDS by Bassett and Mhloyi (1991)). Men benefited most from this volatile and lucrative trade, though some women also took part.

This was to be expected given the general expectation in Buganda that the woman's role is to provide carbohydrate for the household through farm work, whereas the man's role is to provide protein and consumer goods through labour, cash-crop sales and other activities in the extra-village market. Tadria has described this as two circuits of economic activity that intersect within the village and the household (Tadria, 1985, pp. 63, 73 and 92).

Tadria has explored the place of women in Ganda society at some length. In particular, she has looked at the question of the differential access to resources of men and women in this part of Uganda. Locating the origin of some of the present-day inequalities in development in the colonial and post-colonial economy associated with cash-cropping and taxation, and recognizing that 'today both men and women are affected by landlessness' (Tadria, 1985, p. 25), she forms the view that it is women who suffer most. Her survey of 100 households demonstrates this clearly. Most of the poorer households in her sample were female-headed and had land shortages. This is evident from Table 5.1.

The origins of this inequality lie both in the pre-colonial society and in the ways in which colonialism affected that society. It is clear from the observations of the nineteenth-century missionary and writer, Roscoe (1911), that in Buganda women had always been dependent on men for access to land. Thus he says:

> In Uganda the garden and its cultivation have always been the woman's department. Princesses and peasant women alike looked upon cultivation as their special work; the garden with its produce was essentially the wife's domain . . . No woman would remain with a man who did not *give her* a garden and a hoe to dig it with . . . When a man married *he sought a plot of land for his wife* in order that she might settle to work and provide food for the household. A chief had an abundance of land which *he could give to his wife* . . . a peasant, however, had to obtain a plot of land from the King or some chief. (Roscoe, 1911, p. 426. Italics added.)

Within the 'feudal' system of pre-colonial and early colonial Buganda, women gained access to land only by occupying a status *vis-à-vis* men.

Table 5.1 Land distribution by socioeconomic status and gender in a Ganda village

	Land area in acres						
peasant group	no land	<0.25	0.25–<1	1–<2	2–<3	>4	Total
rich							
men			1	3	10	4	18
women	2						2
middle							
men	3		3	8	6		19
women	5		3	3			11
poor							
men	7	5	3	6			21
women	22	1	6				29
Total	40	6	15	20	16	4	100

Source: Tadria, 1985, p. 25.
Note: There are arithmetic mistakes in this table: they appear in the original.

There were a few exceptions to this rule concerning the special rights of women from the royal household (West, 1972, p. 107) but in general this was the case. Indeed, on the death of a man, his land reverted to his clan. His wife had no expectation of continuing her occupation and use of it, once more becoming dependent on the men of her own clan for access to the means of subsistence for herself and her children.

With the ***mailo*** land-tenure reform, this situation changed in principle. West suggests that, after the reform, women were able to gain access by purchasing a tenancy or by inheritance. By the middle of the present century, almost 50 per cent of all inheritors who were registered were women. However, the situation remained fluid and today it is unlikely that the majority of land transactions are formally registered. On the death of a spouse, a widow does not automatically inherit her husband's land. In most cases she inherits personal property while the land reverts to her husband's clan (West, 1972, p. 110). Some women do purchase land; Tadria suggests that one of the effects of the ***mailo*** reforms was to make this option more available to those few women who had sufficient cash income to invest in this way (Tadria, 1985, pp. 101–2).

From observation of the circumstances in present day Buganda, it appears that many women would like to own land but most lack the resources with which to purchase it. The general male opinion is that women should not own land because such economic independence presents a threat to the stability of marriage; but the general opinion of female respondents in our survey was that women should own land, precisely because of the instability of Ganda marriages, a point also remarked by West (West, 1972, p. 109). Some women do own land, obtaining it through purchase or through inheritance, but it remains the case from pre-colonial times that most women have only use-rights obtained through their relationship with a man, either through marriage or through consanguinity. In so far as most women do not have access to sufficient cash to make their own purchases, their main source is through inheritance. Given this and the widespread male attitude that it is not right for women to own land, it must be concluded that few women actually have independent *de facto* rights in land.

This has important consequences. Tadria notes the two overlapping but distinct economic niches occupied by men and by women: men are concerned with cash crops and the market, women with food crops and the domestic sphere. In such circumstances, it was inevitable that a radical inflation of men's cash incomes as compared with those of women would increase the economic insecurity of women and with it their dependence on men for cash and consumer goods. This would result in a change in the relations between the sexes, a change in the nature of what Kandyoti has called 'the patriarchal bargain' (Kandyoti, 1988). This is exactly what seems to have happened in this part of Buganda in the late 1970s and early 1980s as many men became richer through their participation in the ***magendo*** economy. The mechanism in which this further imbalance in the relation between men and women was expressed was through the nature of sexual liaisons.

In Buganda (as in all parts of the world) there are many ways in which men and women may organize their sexual relationships. These include formally defined marriage, both monogamous and polygynous, long-term relationships that do not have *de jure* recognition, casual liaisons and commercial prostitution. In each type, there is an economic core – women can be seen as exchanging sexual favours and often also reproductive potential for economic benefits. The economic transaction may not be the main or express aspect of the relationship for the participants, but given women's underlying unequal access to economic resources, sexual favours and reproductive potential are powerful resources – sometimes the only resources – on their side of the transaction.

During the Amin period, the existence of illicit economic activity in the area had a serious destabilizing effect on the already unequal balance between men and women. Those men who participated in the trade were in receipt of grossly inflated incomes. Their purchasing power grew greatly as compared to that of women who remained locked into the economic circuit of the village economy and within this were largely excluded from the economic security of owning land. Women's landlessness and the general insecurity of title (Tadria, 1985, p. 80) meant that they could not share directly in the economic boom, particularly as the main cash crop, coffee, remained largely in men's hands. One of the few ways in which they could gain access to the new cash and goods appearing in the system was through sexual relationships. This was reinforced by a male ideology which here, as in many other societies, evaluates male status in relation to other men at least partly in terms of sexual conquest and the procreation of children – an ideological association between power, wealth, sexuality and fecundity.

In communities where people's sexual attitudes were already fairly relaxed by the public standards of many other cultures (Kyewalyanga, 1976, pp. 155 and 208), younger women responded to this situation in a number of ways. Some became prostitutes, selling sexual services in return for cash. Others became kept women in stable non-resident relationships where a visiting man gave gifts to help with the household requirements. Yet others entered into various types of marriage. In brief, women gained access to economic resources through a range of sexual relationships with men. This general conclusion about the lack of economic security for women, partly brought about by the exigencies of the colonial economy, was also reached by Bassett and Mhloyi (1991) with reference to Zimbabwe.

The distribution of these types of relationship varied from place to place. Along the main transport routes from the lakeside, 'kept women' and prostitution were to be found. It is likely that opportunities for the more casual types of sexual relationships increased in the 1970s and perhaps encouraged greater 'disassortative' mixing between different sexual-activity groups. The brief case-studies of Kasensero and Kasesere presented below describe socioeconomic circumstances different from those that were directly associated with the peak of the ***magendo*** period, but which produce the same environment where there can be free mixing of sexual partners from what are probably different sexual activity groups and communities. In these two locations, women set up independent households, supporting themselves through petty commodity trading, the provision of services and the maintenance of one or more regular lovers who help them financially. In such circumstances, the rate of partner change can be assumed to be fairly rapid. In the more remote villages, more traditional relationships based on beer-drinking and sex seem to be the norm, but it was not until AIDS had been introduced by travelling men, often of a higher educational status or with cash to spend, that these traditional networks took the infection further into the rural areas. It is tempting to hypothesize that this difference in pattern between the lakeside and the inland villages accounts for a slower spread of the pandemic in such sub-counties as Kagamba, Byakabanda and Lwamagwa, where the traditional sexual networks became infected at a later date (see Figure 2.7, Chapter 2). The importance of these observations is that, on the foundation of already unequal gender relations, the ***magendo*** system altered the market balance between men and women even further to the advantage of men.

Ganda marriage: social identity and property relationships

The majority of women in Buganda are married. Marriage establishes adult status for men and for women. Traditionally, a Ganda man could not be described as a homeowner (***semakka***) unless he had a wife; a Ganda woman remained a girl (***muwala***) in the eyes of the community if she did not have her own banana grove and kitchen. A man was expected to offer a banana grove (***lusuku***) to his wife or help her establish one soon after marriage. Men owned the land and women had usufructory rights. Since the 1900 Agreement, which, among other things, introduced the system of individualized land tenure (***mailo***) women have been preoccupied with acquiring ownership of land rather than merely having access to it as mothers and wives. Wives and sisters of chiefs were the first to acquire ***mailo*** land. But as cotton and coffee production boomed in the 1950s and 1960s ordinary women, too, bought land. This increased and gave new focus to the already existing tensions between men and women. Women were demanding more consumer goods from the husbands who appropriated cash-crop money and they were demanding more land to cultivate in order to have their own independent sources of cash. Women were threatening to stop cultivating for their husbands and to establish and support themselves independently. In the 1950s and early 1960s desertions (***okunoba***) by wives became a common phenomenon as women either ran off with rich men (because 'My husband did not give me enough land to cultivate,' i.e. he was not looking after me), or simply ran off with a lover, or established themselves as independent householders (***nakyeyombekeddesing***).

Desertion became a safety valve for women who were legally denied recourse to the divorce of adulterous, mean or authoritarian husbands, whereas the husbands could divorce their wives for adultery or even bad cooking. Women who deserted their husbands went to far off places; or, if they returned to their brothers' or parents' homes, they had to be discreet with any amorous involvements. Although Ganda bride wealth was a token, deserted husbands, if they had money, resorted to hiring 'adultery detectives'. When discovered, the woman and her lover would be taken to court and fined. Then the matter would rest. The result of this marital instability is that Ganda children experience many different rearing and living arrangements. When a woman deserts, her older children stay with their father and are thus raised by a step-mother or several step-mothers if the father marries a succession of wives who decide to desert him. The mother takes her younger children with her and they may be reared by her brother and his wife, or a sister and her husband or by her mother and father. Women with no close, continuing contacts with their immediate family or who are very young may entrust their children to a close friend, maternal uncle or paternal aunt. One observer of these arrangements commented that 'the Ganda seem to exchange children like Westerners do Christmas cards'.[1]

In the 1950s and 1960s the urban areas offered rural women alternative ways of earning a living which would liberate them from the constraints of landlessness and poverty. The majority of these female urban self-employed migrants saved money and bought land-occupancy rights (***kibanja***) or actually succeeded in acquiring ***mailo*** land. Women bequeathed land to their daughters and sisters. The majority of women, however, still had only usufructory land rights, but widows stayed on the land until they remarried. Most widows preferred not to remarry because widowhood (***namwandu***) had become a prestigious female status with its implications of independence and freedom. At this time some educated, Christian men who had remarried started bequeathing land to their wives and daughters. But women's control of land was and remains tenuous, as in cases where women fear that their husbands' grown-up sons might dispossess their step-mother of the land.

Women and change: the 1960s

The 1960s were years of expanding economic opportunities. For the educated, the sky was the limit, while for the uneducated coffee production and informal sector service activities provided respectable livelihoods. Many women strove for economic and ultimately social independence from men. At the same time an income redistributive system developed as women devised ways to share the wealth of the predominantly male group who enjoyed more privileges in education and employment. Some women were desperate – they were either unemployed or earned a pittance from their jobs or from self-employed activities such as handicraft production and brewing – and sometimes turned to prostitution. Some educated women needed to supplement the incomes from their lower strata professional jobs. These women formed liaisons with powerful men. There were other women whose passport to men's hearts and wealth was their beauty and men were known to neglect or even disinherit their families over them. The prostitutes used sex as a direct way to getting money. The concubines or 'kept women' used sex and other social accomplishments to gain access to resources such as houses, land, jobs or cars and even marriage; they often received payment in kind in the form of expensive luxuries. Needless to say, there were successful and unsuccessful prostitutes and concubines, so standards of wealth and achievement varied depending on their beauty and social skills. The majority of women who were in fact married were often preoccupied with the unmarried women whom they despised as 'purse snatchers'. The men for their part denounced prostitutes who were seen as lazy town-women who should be engaged in productive agricultural labour in the rural areas. Some prostitutes said they had migrated to towns to escape either rural poverty or back-breaking agricultural work. Law makers (who were usually male) rejected élite women's efforts to obtain legislation guaranteeing and protecting women's property rights and securing inheritance rights for their children.

From the mid-1970s some women traders also participated in the ***magendo*** trade in their own right or as surrogates for the military officers who were otherwise occupied with misusing power. Some female house servants and urban school children rebelled against employers and guardians to engage in petty trading. Meanwhile a new class consisting of civilians and soldiers had acquired the property of the expelled Asians. These ***mafuta mingi*** (literally 'a lot of fat') displayed their conspicuous consumption through spending lavishly on women, so they attracted the so-called ***mafutaminglets*** (a term that might be translated as '***magendo*** groupies').

Civil unrest and warfare: the last two decades

While the general civil disorder that had characterized the Amin years and the second term of office of Milton Obote (1979–86) disrupted social relations in this area, episodes of full-scale warfare added another element. Both the 'Liberation War' that overthrew Amin in 1979 and the civil war that brought the government of President Yoweri Museveni to power had major effects. The combination of an invading army moving through the area from the Tanzanian border, together with fatalities among local men, meant there were additional opportunities for the disease to spread (Giller et al., 1991). Apart from rape and some localized shortages of men, times of war have always been times of decreased sexual control. Certainly some of our informants reported this to have been the case (Nalwanga-Sebina, 1989, p. 19). During times of war: 'soldiers took money and sometimes killed men, but always took money and sex from women', said one informant. In general, then, Rakai people suffered loss of property and lives during the 1979 Liberation War in which the invading exiles supported by the Tanzanian army defeated Amin's troops. People say of this time: 'The Tanzanian soldiers did save a lot of women from rape, and did not loot. It was local people who did most of that.'

Thus ***magendo***, the changing position of women, warfare and civil unrest all combined to make Rakai into a risk focus. The nature of the high-risk environments created by these circumstances can be seen from an examination of the two communities of Kasensero and Kasesere as they appeared in 1989–90.

Kasensero and Kasesere

Kasensero

This village serves as an example of the way in which many of the factors outlined above have come together to create a focus of extremely high risk within Rakai District. It is located on the shore of Lake Victoria about five kilometres from the Tanzanian border to the south, and is joined to other centres inland only by a track, at present in a near terminal state of disrepair, which stretches across the marshes for its last four kilometres. The main economic activities are fishing and trade, both legal and illegal, and the track carries a considerable amount of slow-moving traffic carrying fish and other merchandise. The provision of eating and sleeping facilities for the men who are involved in these activities seemed to be the main occupation of the women in the village. Fishing is organized in the following way. A man will accumulate enough capital to purchase or build a boat. He may do this by working for somebody else, or through smuggling. He will then find a crew of younger men, typically in the mid to late teens and early twenties. The crew will jointly take a 30 per cent share of the catch, the remainder accruing to the boat owner. Given the number of households in the inland villages of Rakai who reported that young male relatives had left the village to live by the lakeside, this appears to be an attractive proposition. In addition, the endless stream of fish traders making their way laboriously from the lakeside, each wheeling a single, huge Nile perch in a basket to markets as far as Kyotera 30 kilometres away, indicates that there is considerable demand for the fish. Fishing is combined with smuggling, making the village a magnet for young men who want economic independence and an alternative to the farm.

The majority of the people living in Kasensero are less than 30 years old. The women's economic activities have to do with servicing the men. The main street is lined with 'lodges' which provide sleeping accommodation and some food, as well as small eating places. The women who run these places are either unmarried or live separately from their husbands. In interviews, many of them reported that they had left relationships in other parts of Rakai and Uganda as well as in Tanzania and that they had made their way to Kasensero in order to find the economic independence and security unavailable elsewhere. Some brief histories will serve to illustrate the quality of their lives.

Zeinab, aged 30, runs a lodge by the customs house. She has a male companion who contributes to her household, but she does not see this as a permanent relationship. She has three children aged between 10 and 14 who live with her husband in Kampala. She has lived in Kasensero for nine years. Her aim in coming to the village is to save enough money to buy some land in her home village. She has no other way of getting land, as her father (a Muslim) left his land to her brothers and they have refused to share it with her.

Catherine has been in Kasensero for two years. She was born at Kyebe about 20 kilometres inland from Kasensero. She has had a number of stable relationships both in Uganda and in Tanzania. She has four children, two of whom live with her first husband in Tanzania. She left her second husband after four months and the child of that marriage is with her. Catherine supports herself by selling cooked food; she sends some of her income to her mother who looks after her young daughter. Her mother and brother are farmers and have some surplus. From this they gave her 500 shillings to start her business. She is saving to purchase land in her home village. She worries a lot about AIDS, particularly because she has her young teenage daughter living with her in Kasensero.

Two further vignettes say something of the general situation of the men in this village.

John is 17 years old. His father is a farmer aged 53 and lives in a village about ten miles away. His mother died a year or two ago. He has some primary school education and used to work on his father's farm. However, as there was no cash available for him from this, last year he decided to come to Kasensero to work as a fisherman. He is a member of the crew of a boat owned by a man he met when he came here. He would not say what his income was – 'it depends on the catch' – but he had all the external accoutrements of cash and of male style in this place – a wrist-watch, a smart jacket and sunglasses. He was not concerned about catching AIDS, but spoke of the fact that few new boats were being built 'because it takes five years to recoup the investment': the clear implication was that few men expected to live long enough to profit from such a long-term venture. His aim in coming to the settlement is not to accumulate money for any specific purpose, although he would like to get married – 'but not yet'.

Mark is aged 22. His father is a farmer near Kyotera. He left home five years ago to earn money as a fisherman. He works as a member of the crew, but because of his knowledge and experience he takes a share of the owner's portion of the proceeds of each catch. He also does some smuggling, in particular motor-car tyres from Uganda to Tanzania; this brings in more money than the fishing. He has a permanent relationship with a woman who sells prepared food in the village and they have a baby. He also has another child in his home village.

The general impression of this community is one of rootless people who have come here to earn cash by whatever means presents itself. There were few formal marriages and among such a young population, the rate of partner change must be assumed to be quite rapid.

It is in such circumstances that we see the social and economic milieu that produces an area of high risk of disease transmission. Certainly many of the men to whom we spoke gave the impression they did not expect to live for very long. They knew that many people in the village had already died from AIDS. There was a degree of bravado and machismo associated with this – fishing and smuggling are high-risk activities, what difference does an additional risk make? In contrast, the women seemed to take a longer view, although they were also very aware of the dangers of the disease.

Kasesere

That fishing, trading and farming are interlinked occupations within which people circulate between the very high-risk areas into the surrounding countryside is illustrated by this account of Kasesere fishing village on an inland lake, Kijane Balola, near the Rakai District headquarters. The village is established on public land but one of the older residents reported that he had a traditional claim to distributing the land to deserving villagers to build huts or grow food. He owns three huts and generally acts as the authority figure in the community. He is the chairman of the local Resistance Council. The population of this settlement varies between 40 and 150 people during the high fishing peaks. There are ten house owners renting out simple mud and wattle papyrus thatched rooms which are generally shared by a number of people. About three-quarters of the residents are single young men who divide the year between the seasonal fishing peaks, which they spend in Kasesere and the agricultural peaks, which they spend in their home villages. Some of the fishermen drift off after some years and find more interesting or profitable work than fishing. Others, after working for some years in the fishing village, may build their own small houses and even buy their own boats and nets. There is a type of 'career' structure: the more these men become established in the life of the settlement, the less they visit their home villages. One general-store manager came as a hired hand but after a time opened a shop and brought his wife who wanted to escape village gossip about their 12-year-old childless marriage.

There are about eight categories in this settlement. These are:

(1) five boat owners who hire people to work for them and also hire out boats and nets;
(2) fishermen boat owners;
(3) fishermen boat hirers;
(4) hired fishermen;
(5) odd job men and (old) women who string fish (***nkejje***), fetch water and firewood and wash clothes for others;
(6) four shop owners, and tailors;
(7) beer brewers, gin distillers and purveyors;
(8) young women who come because of relationships with men or because they hope to find a job. If they fail to find a semi-attachment to a man or to work for a bar owner, such young women usually leave the settlement. However, they are soon replaced.

There was one young man at the settlement who was ill. People said he probably had AIDS. He had no family elsewhere. Ten men and five women who left and never returned were known to have died of AIDS. The oldest woman, in her fifties, who is one of the earliest remaining settlers, remembers each one and says she grieves for them, their families and anyone with whom they may have come in contact along the fish-trade route.

Conclusion

This chapter has shown through historical and contemporary data how an interlocked set of economic and political processes led to the development of high-risk foci in which AIDS could manifest itself in the particular way that it has in Buganda and in Rakai District. The smuggling is now much reduced, but the character of these high-risk villages has not changed. The case-studies of Kasensero and Kasesere suggest the circumstances in which economic insecurity for women and their attempts to cope with this through various types of liaison with men, create the social relations characteristic of a focus of infection in this environment. It has to be emphasized that these are not typical Ganda villages. However, because of their particular place in the economic and social history of Uganda, Buganda and Rakai, they are important foci of infection and epitomes of the types of process described in this chapter.

Note

1 Personal communication from Dr Christine Obbo.

A note on further reading

Further reading on the background to Uganda can be found in Wiebe and Dodge, 1987; Southall, 1988; Sathyamurthy, 1986; Mamdani, 1975; Hansen and Twaddle, 1988. On the historical development of gender relations and their implications for AIDS, see Kisekka and Otesanya, 1988, and inevitably partial reviews of sexual behaviour in Africa by Standing and Kisekka, 1989, and by Larson, 1989.

6

How Households, Families and Communities Cope with AIDS

This chapter examines the ways in which AIDS impacts upon households, families and communities, and describes the main coping mechanisms that have been developed in the worst affected areas of Rakai District. The main fieldwork sites were the settlements of Guanda, Kasensero, Kyebe, Kiembe, Kooki and Luanda (January–March 1989) and Luanda (June 1989–January 1990).

Categories of unaffected, AIDS afflicted and AIDS affected households

It is useful to expand upon the three broad categories of household in any community with AIDS as we distinguished them in Chapter 4. *Unaffected* households are those in which no member is ill or has died from AIDS and which has not been affected by the illness or death of a member of any related household. The notion of being unaffected can be specified more precisely as describing a situation in which no additional burden, either of time or economic or financial resources, has accrued to a member of this household. However this is a relative term. Most households are in some sense affected in a milieu in which the disease is so widespread.

In AIDS *afflicted* households, the impact of the epidemic has been direct. A member of the household is either ill or has died from the disease. Resources have to be reallocated in order to deal with the problem. Finally, AIDS *affected* households are those that have been affected by the disease either through the death of a family (not necessarily household) member who was contributing cash, labour or other support, or because the death or illness of a family member has meant that, for example, orphans have come to join the household. These are all events that place additional demands on existing resources. There are also households which are both AIDS afflicted and AIDS affected.

Each of the latter two types of household circumstance will require different strategies for coping with the downstream impact of the AIDS pandemic. There will obviously be differences within these types and they will also reflect other socioeconomic and demographic variables discussed in the next subsections.

Different households with different coping mechanisms

First, the cumulative impact of one, two or more AIDS deaths upon the household through time is examined. Obviously, increasing numbers of deaths or orphans received from afflicted households through time will call for more radical and far-reaching coping mechanisms. Later subsections examine the changes in household structure in Buganda and the importance of the point at which AIDS afflicts or affects a household in its 'normal' demographic cycle. The household is formed usually with a young couple; then children are born, grow up and leave to form new households; and so the demographic profile (and therefore labour availability) of a household is constantly changing. Below we describe the part played by socioeconomic differentiation on the impact of AIDS – some households have more social and economic resources than others to adapt to and mitigate the effects of sickness and loss of household members, or to care for orphans. A set of case-studies follow. In these we cross-classify the households by their position in the 'normal' demographic cycle by socioeconomic status and by their responses to the effects of AIDS. This section ends with some simple quantitative analysis of the different responses of households in different categories of fall or rise in the ratio of producers to consumers, together with their AIDS status. The final section of this chapter summarizes, in list and diagrammatic form, the major household-coping mechanisms.

The impact of AIDS upon the household through time

As illness and death in a family or household occurs once, twice or more, the impact is felt more strongly, demanding new and ever more radical coping strategies. As we have suggested, the impact of AIDS deaths upon the household is unlike that of other disasters such as drought, famines, earthquakes or even war. It is gradual and incremental and occurs over a period of at least five years. Thus the onset of the symptoms of full-blown AIDS, increasing debility and finally death, afflicts one, two, three, and sometimes up to eight household members, as was found in one household in our field survey in Rakai District.

The impact will probably be unevenly distributed between households, since AIDS deaths tend to be clustered within households for two reasons. First, there is a high risk of inter-spouse infection; secondly there are probably distinctive behaviour patterns between households (based on socioeconomic status) which significantly increase the risk of infection.

This is particularly the case with households which include (male) traders who travel away from home for long periods, (female) beer-sellers and bartenders and men and women who have received more than primary education.

In this section an idealized longitudinal model of interrelated and incremental impact is described. These impacts called for a variety of coping mechanisms that are brought into action by the household. All these are brought about by illness followed by the death of one or more members of the household.

These mechanisms are effected at different stages of a household's experience of the AIDS pandemic and therefore they occur in an ordered sequence. To illustrate this, an idealized longitudinal model is described below in Figure 6.1. This is derived from our interviews in Rakai District carried out in January to March 1989. It is a composite description (although not an extreme one) of the varied experience of families in the District.

Stage 1 (notional date, 1980). Here is a household consisting of a man, a woman, two daughters of twelve and ten, and three sons of five, three and one, with two older sons aged seventeen and eighteen, who do not live in the household, but send occasional cash remittances from their trading and fishing activities. The household cultivates two acres, growing banana, the main staple in this area, beans, Irish potato, sweet potato, groundnuts, some kitchen vegetables and two-thirds of an acre of coffee; the latter is the main cash crop. They employ two labourers for one and a half months each for weeding and planting the annual crops. They use small amounts of herbicide and pesticide on the farm.

Stage 2 (1983). The fisherman son dies of AIDS and the other falls sick and is unable to trade. Remittances cease but the household's labour force remains intact. The agricultural labourers can no longer be hired. Cash shortages force a stop to the use of herbicides (increasing the demand for labour at existing production levels) and pesticides (reducing average yields and the stability of yields over the following years). Some increase in the working day on the farm by household members becomes necessary.

Stage 3 (1985). The husband falls sick from AIDS and the second trading son dies. The household cannot increase its working day, since the husband can no longer work, and his wife has to spend time caring for him. The youngest girl (now 15) is taken out of school to reduce pressure on the household budget, to help on the farm and to care for her father. Coffee is neglected and cash from its sale is reduced. The area under cassava, a crop relatively undemanding of labour, is increased, while beans, groundnuts and Irish potatoes are reduced. The banana garden is neglected, but for a year or two this is relatively forgiving of neglect. Extra expense is incurred in a search for a 'cure' for the husband.

Stage 4 (1987). The husband dies, and the wife becomes ill with AIDS. Funeral expenses for the husband are incurred to the cost of 80,000 shillings (about £85 or $140 at the parallel rate of exchange). All the remaining

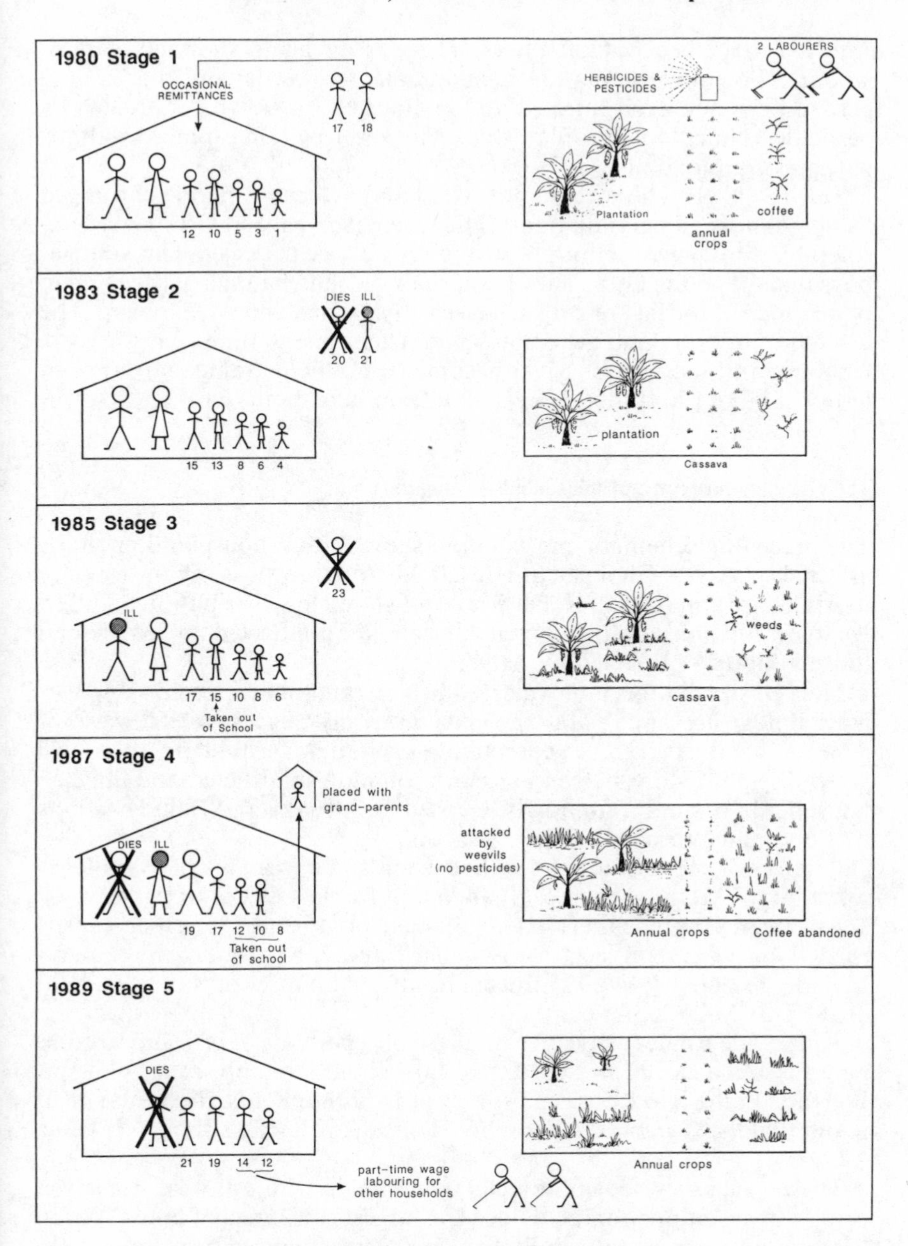

Figure 6.1 Incremental AIDS deaths within a household

Source: Authors' own fieldwork.

children are taken out of school. There is an acute shortage of farm labour. The coffee plantation is abandoned to weeds, and the banana garden becomes weed infested and is attacked by weevils, untreated by pesticides. Nutritional standards fall. The youngest son is placed with his paternal grandparents.

Stage 5 (1989). The wife dies of AIDS and a cheap funeral is arranged by neighbours and grandparents. The seventeen-year-old girl spends considerable time away selling beer and trying to earn cash. The younger boys remain on the farm. They manage to keep the banana garden and to plant crops of the last resort such as cassava, yams and sweet potato. The two eldest boys try and get employment from time to time near the road. Clothing and food for survival becomes a problem. Some assistance in terms of labour and food is available from neighbours.

The changing structure of households in Buganda

The preceding schematic presentation shows how a household might be affected by AIDS. Changes in household structure, possibly in response to AIDS, are apparent in Rakai. Table 6.1 shows the distribution of different types of household between households unaffected by AIDS and those afflicted or affected by AIDS.

The cell sizes in this table (derived from a randomly selected sample of households) are small and the data are only suggestive. However it appears that the two/three generation + orphan household and the single person living by him/herself are more common in afflicted and affected households. A similar finding is reported for Masaka and Rakai Districts by Bond and Vincent (1990).

In terms of overall numbers of households in our survey of the villages worst hit by AIDS – Kiembe, Guanda and Kooki – 14 out of a total of 69 (20 per cent) were either AIDS afflicted or affected. Another census carried out by the project, of Kakooma 1 and 2, Kitemba, Kituento and Luanda, revealed 49 AIDS afflicted or affected households out of a total of 185 surveyed (26 per cent).

To provide some perspective on these observations, it is useful to compare these data with an analysis of household structure in a village in Buganda in the late 1950s. In a survey of 165 households, Richards (1966) found that the structure of households was distributed as shown in Table 6.2.

These categories are not directly comparable with our own. However, certain observations can be made. The largest categories of household in the table are couples with children (37 per cent) followed by couples without children (21 per cent). This contrasts sharply with our own data in Table 6.1 where couples without children ('couple') make up only five per cent of households in the unaffected Ganda households in our sample. Richards does not note any polygamous co-resident households, whereas these make up nearly eight per cent of the total in our material. Perhaps

Table 6.1 Types of household structures: Kiembe, Guanda, Kooki villages, Rakai District, 1989

Type of household	Unaffected households		AIDS afflicted or affected households	
	number	**percent**	**number**	**percent**
Single person	0	0	1	7
Couple	3	5.5	1	7
2 generation*	40	73	8	57
Polygynous with male in residence	4	8	0	0
3 generation	2	3.5	1	7
2/3 generation plus additional members (orphans)	5	9	3	21.5
Other	1	2	0	0
Total †	55	101.5	14	99.5

* Including single-person households, usually female-headed but occasionally male-headed, and polygynous households where the man may not always be in residence.

† Totals do not sum to 100% as cell values have been rounded to take account of the small sample size.

Source: Authors' own fieldwork.

the most significant difference between the two sets of data is the absence of any mention of orphans in Richards' data. She was not interested in the problem and so does not categorize them separately. There is also no mention of three generation households where grandparents are looking after children. These may have been included within her category 'couples with children'. If this were so, then we might have expected that category to be larger than it is.

We must therefore tentatively conclude that the presence of a significant percentage of households in which grandparents care for orphans is a new development, at least in part a response to the pandemic. Also of significance is the number of households in the category 'couples without children'. We can assume that the majority of these would have been newly married couples. They make up 21 per cent of the households in Richards' sample. By contrast, in our own sample, they make up only five per cent in the unaffected households. The possibility must be considered that this indicates that in Buganda today young people are hesitating to marry and hence that there are fewer newly married couples in the sample. That this may be the case is supported by anecdotal reports that there are

Table 6.2 Household types in a Ganda village in the late 1950s

	no.	%
Couples living with children	60	36
Couples without children	35	21
Men living alone	25	15
Men living with an older son or daughter	3	2
Women living alone	9	5
Women living with children	11	7
Women living with an adult relative	4	2
Couples living with an adult relative	11	7
Others	7	4
Total	165	100

Source: Richards, 1966, p. 71.

now very few marriages taking place in Rakai District,[1] as well as by data from our own survey that suggests that young people are delaying marriage. It is likely that in all the villages in Rakai District we surveyed, there were fewer children than would be expected from a stable population in a country like Uganda. Hunter (Hunter and Dunn 1989) gives the percentage for children from 0–4 years as being 15 per cent and 16 per cent for two of the worst infected sub-counties. This is almost identical to the percentages of female and male children of the same age group in our own survey (see Table 2.6). Hunter estimates that the proportion of children in this age group in a population like Uganda should be 22 per cent.

Although we consider that this is a high estimate, we can conclude that the percentage of children of this age group found in our and Hunter's villages is probably lower than expected. The reasons for this may well include higher mortality because of the vertical transmission of the disease to children; the raised mortality of very young orphans; and possible under-recording of the existence of children because of the migration of very young children with their mothers who have deserted sick husbands or who have migrated. It has also been suggested by some of the people from Rakai that the lack of children is due to increased celibacy. We met cases where mothers have strongly counselled their children not to marry; a local midwife, one of our informants, said that she was almost without work because of the lack of births. Some women reported growing tensions between spouses over the problem of unwanted sex where it was suspected that a male partner might already be infected.

The 'normal' demographic cycle in Ganda households

In 'normal' times, households grow with the birth of children, its members grow older and the children grow up and leave the parental home.

The point at which AIDS intervenes in this normal process is significant for the impact on that household and the responses it is able to make. A simple typology might identify households as 'young', with the household head (male or female) aged less than 30, 'mature, with a household head between 30–50 years', and 'declining' with a household head over 55. However, this assumption of household structure being linked reliably to the age of the household head, deriving as it does from the work of Chayanov on the Russian peasant household (Thorner et al., 1966), has to be adapted in the Buganda context, given the existence of polygyny and the presence of a large number of female-headed households. A feature of Buganda is the presence of ***nakyeyombekedde***, 'women who have built for themselves', meaning women in semi-independent relationships with men, who have often been married or have lived in a male-female household but now live alone with their children. Indeed, Tadria found that 42 per cent of the households in her sample were female-headed (Tadria, 1985, p. 27). In addition, men in their fifties or older who are fairly wealthy may marry again and rear a young family. In such cases, some of the characteristics of a 'young' household will reappear although the man who is considered to be the household head is in late middle age.

Social differentiation

Figure 4.2 in Chapter 4 has suggested how illness and death may affect access to resources and the level and type of vulnerability in different types of household. In box (1), an array of *households* is shown with different *resources*, both economic and social. The resources most relevant to coping with AIDS in the household are labour (differentiated by gender since men and women perform different tasks); land; cash reserves; household and family skills (caring, parenting, managing the household in the face of crisis); income-earning activities (farming, trading, fishing, government employment and teaching, etc.); and the wealth of close relatives who may be able to help care for orphans, lending money, etc. However, equally as important as *economic resources*, there are *social resources*, which are networks of material and emotional support that may have to be called upon to mobilize these resources. These are also of great importance in the coping process. In the next section we show by means of detailed case-studies how access to resources is also affected by the position of a household in its demographic cycle and its socioeconomic position and how these influence the household's response to the impact of the disease.

Case-studies of the impact of AIDS on households of different demographic and economic status[2]

A simple cross-classification of demographic and socioeconomic status is used. We can divide households into the three broad categories rich, middle and poor. These are, of course, relative terms. In practice, we distinguished them simply as follows:

Rich: these households had some or all of the following attributes: adequate or surplus land, ability to hire labour, involvement in trade or other non-agricultural occupation;

Middle: these households had non-agricultural sources of income, occasionally employed labour and had adequate land or sufficient non-agricultural income to be able to rent land or purchase food;

Poor: these households were short of land, experienced labour shortages which could not be met by hiring, and had few if any sources of cash income; or alternatively, their land resources were so low as to require that household labour was hired out to other households.

These three categories are cross-classified with the three normal household cycle categories – 'young', 'middle' and 'declining' – described above, and with AIDS status. Cross-classification of these three parameters produces 18 conceptual types. These are summarized in Table 6.3. Some of these types are then illustrated by case material from our fieldwork. These illustrative types are indicated by an asterisk in the Table. The fact that we do not provide case-studies of the other types indicates that we did not find households with such combinations of characteristics in our survey. Needless to say, here, as throughout this book, all the names of the people concerned have been changed.

Case-studies of household responses to AIDS

Type 3: A rich AIDS-affected household in decline

This household consists of a man aged 80 and his two wives (his second and fourth) aged 64 and 28. The first and third wives died in 1977 leaving behind three children. One of the daughters,now aged 55, is divorced and has come back to stay with her father. The second wife had eight children of whom six are living. Two of her sons died (both from AIDS) leaving behind nine orphans each – all of whom stay with this old man and his two wives.

There is a large labour force as not all the children are too young to work on the farm, although none of them is beyond mid-teens. They cultivate bananas, potatoes, groundnuts and yams. This household also owns cattle and uses manure to improve yields. They sell some of their surplus bananas.

This is an example of a wealthy household. It used to employ a lot of labour in the 1960s and 1970s and has considerable assets. Despite the deaths of the man's children, the household can cope by employing its wealth and still-intact labour force. In addition, the man has educated his

Table 6.3 Types of households in an AIDS-infected area of Uganda

Household type	AIDS status afffected (A) afflicted (B)	social status rich = R middle = M poor = P	developmental cycle young = Y mature = M in decline = D
1	A	R	Y
2			M
3*			D
4	B	R	Y
5*			M
6*			D
7*	A	M	Y
8			M
9			D
10	B	M	Y
11*			M
12*			D
13	A	P	Y
14			M
15			D
16	B	P	Y
17*			M
18*			D

Asterisked categories are illustrated by case-studies in the text

older children; one of them is now a headmaster who provides financial support for his dead brothers' children while the old man's land and the household labour provides their subsistence.

Type 5: A rich AIDS-afflicted mature household

This household consists of a man aged 36, a woman aged 32, their daughter of 18 and two other children aged 12 and five. The two younger children are in school. There is one other child aged less than five years.

They have two farm plots, one small one in this village, another in a nearby village. It is almost certain that this man had another wife in the latter village. They grow beans, cassava, potatoes, maize, groundnuts, sorghum and bananas. They are not yet harvesting coffee but are in the process of establishing a coffee plantation in their other plot. The household employs labour for land preparation.

They sell considerable amounts of their produce, estimating the proportions of their production sent to market as follows: beans 30 per cent;

maize 80 per cent; groundnuts 30 per cent. Recently, their teenage son died from AIDS. He was a fisherman.

The husband also used to be a fisherman, then a trader; now he has turned to farming. He left the fishing to his son and ceased trading because it was unprofitable.

This household has experienced the recent loss of an income source through the son's death. But this does not seem to have made it particularly vulnerable because there are enough resources, from some continuing trading and fishing, to allow the establishment of a second farm and also to employ labour. Their teenage daughter also contributes labour, so the household has three adult workers and three dependent children. There is some depth of defence in this household, mainly from savings and also from sales of crops. There is also a favourable balance of consumers to workers. Although the death of the son has removed one income source, it has not had a profound effect on the household's ability to cope with its subsistence needs; the children have not been withdrawn from school. Thus this household is only vulnerable to repeated and cumulative deaths from AIDS.

Type 3/Type 6: A rich AIDS-afflicted and AIDS-affected household in decline

This household has coped by combining members of other related households which were living separately until affected and afflicted by the disease.

The man is very old (aged 78) and is married. His wife is also elderly. They had ten children, but a daughter and two sons have died of AIDS. Until the disease struck, this couple was living in an independent household, farming and living off their assets with some help from their son and his wife who lived nearby, as well as hiring farm labour to undertake the heavy cultivation work.

The son died from AIDS in 1986 at the age of 35. His widow brought her three daughters to live with the old couple. One other son, aged 38, has now joined this household. He is ill – probably with AIDS, as his wife died of AIDS within the last year.

The farm work is done in part by the household members, but they also hire Rwandan labourers to plant cassava and cut banana stems. They also grow beans and a range of other annuals, including groundnuts and sorghum. They used to have a coffee plantation but this has been abandoned because of insect infestation.

They now get cash from the sale of bananas and beer as well as from licensed distillation of a local alcohol, ***waragi***.

This household has been severely afflicted and affected by the disease. The household was quite wealthy and still has considerable resources, but, predictably, is now experiencing stress and has taken the three grandchildren out of school in order to relieve the pressures of paying for labour and also to help with the domestic work.

Type 7: Middle income, AIDS-affected, young household

John and Mary, a couple in their thirties, have lost a total of ten relatives in the last two years. They have four children, and they have had to take in six orphans. John has lost three brothers and their wives. Mary has lost a sister with her husband, and an uncle and niece. John's relatives live in Kampala and Masaka District, so visiting them requires major expenditure on transport. Mary has a sister who has taken two orphans in addition to two of her own children; her income comes from selling beer and retailing mats and baskets.

Last year, the couple, who are in minor government jobs, decided to borrow unused land from John's father and grow maize, Irish potatoes and bananas for sale. They say that their salaries are so low they are hardly worth drawing. Rather than hire expensive labourers, Mary has invited her aunt and cousin to come and help with the children, cooking and gardening. There has also been an increase in the working day for most household members.

Type 11: Middle income, AIDS-afflicted, mature household

Jane is aged about 45 and is a widow. Her husband died of AIDS in 1987. She had eight children, four of whom have died, an infant and a daughter of 14, plus one other (none of these deaths was said to be from AIDS) and recently a son aged 25 who did have AIDS. When this son died, his wife became very frightened and ran away, leaving their son aged four in the care of his grandmother. The other children have left home and married. In the household there remains Jane's son aged about 22 (who does not help in the home or the farm at all) and a daughter aged 11, as well as the four-year-old grandchild. There were three fresh graves in the farm.

Jane owns a coffee plot, but it has largely returned to bush because there is no labour available to maintain it. She grows bananas, potatoes and cassava. She raises cash from the sale of mats and baskets, which she makes, as well as from the sale of coffee and sweet potatoes. The main labour input is her own and that of the 11-year-old daughter. But she also hires some labour on a sharecropping basis and this allows some coffee to be cultivated.

This case indicates a household which has just enough resources of labour and also of land to be able to cope quite effectively with the impact of the disease. Having adequate land has enabled this woman to enter sharecropping agreements with landless or land-short people. This is an effective method of coping with agricultural production in cases where there is a land surplus. However, she has changed her cropping pattern, and largely abandoned coffee production in favour of food crops. Her cash income is derived from her handicraft production. She has also taken her 12-year-old daughter out of school. In the medium term, this daughter will contribute more labour on the farm and to the home.

In summary, this household, with its assets of land, has experienced a drop in its level of life, changed its cropping pattern and entered sharecropping arrangements. The household has also been able to take in an additional consumer (the orphaned grandchild). However, given that her husband died from AIDS, it is possible that this woman may also develop the disease. Were this to happen, the result would be a remnant household consisting of her daughter and the grandchild. The son would probably move away; he is already largely resident in Kasensero and she complained that he made no financial or labour contribution to the household.

Type 12: Middle income, AIDS-afflicted household in decline

This household consists of a man, Robert, aged 70, his daughter aged 30 (divorced, her children having left with her ex-husband) and a son aged 15. Robert's first wife left him, as did his second wife. Two children from these marriages, a girl aged 35 and a boy aged 37, died in 1987 from AIDS.

Crops grown are: coffee, bananas, cassava, potatoes, yams and beans. Their income is supplemented by smoking fish and selling it. The plot size is 1.5 acres. This is not all used and they have plans to clear more land in the next cultivation season. They say they would like to employ additional labour but have insufficient cash to do so.

The two children who died from AIDS used to help on the farm; the result of their deaths was that the banana farm and the annual crop land returned to bush. In addition, the 15-year-old son was taken out of school in order to help on the farm.

This household has been affected by the deaths of two of its members – land has returned to bush and a child has been withdrawn from school. The lack of cash means that the cropped area has effectively been reduced, but fish-processing and trading provides a minor source of additional income. This household is quite vulnerable. Were it not for the presence of Robert's divorced daughter (who may remarry) there would doubtless be a labour problem on the farm, given his advanced age.

Type 17: A poor AIDS-afflicted mature household

This is one of the most tragic cases we encountered. It is hard to describe it as a household; perhaps 'remnant household' would be more accurate.

The man lived alone in a bare hut, sleeping on the floor. His possessions appeared to be little beyond a blanket and a pot over a meagre fire upon which he was cooking some bananas. He was said to be 45 years old but looked considerably older. He was clearly very disturbed and could not be interviewed. Information was obtained from others nearby.

Only a few years ago, this was a substantial household with a reasonable farm supplemented by fishing. His wife and eight of his teenage and adult children had died of AIDS within the last few years. He had no relatives living in the village and supported himself by cultivating and selling some of his bananas. Onlookers said of him 'he is not expected to marry again'. This comment from a bystander can be interpreted as follows. It is widely known that his wife died from the disease, therefore it is quite likely that he is himself infected. Were this not so, in any case he now had no assets or resources which would allow him to remarry and establish a new household.

This case illustrates how, in an extreme case, the costs of nursing AIDS victims combined with the disappearance of the family has led to a state of utter poverty where life is sustained at a bare minimum. The entire family-support system has gone; this man was destitute and isolated.

Type 18: A poor (male-headed) AIDS-afflicted household in decline

This is a household in which there have been two deaths in the past few years. Patrick's wife died from 'generalized illness' and his 16-year-old daughter died of AIDS in 1988. Three people remain: Patrick aged 53, a son aged 17 (who is a fisherman at Kasensero and does not work on the farm at all) and a daughter aged 12 (who has been withdrawn from school).

They have little land and they grow coffee, bananas and some sweet potatoes and cassava. There is not enough land to merit the employment of labourers and in any case, they have no money to pay them.

As a result of the deaths, Patrick has lost the daughter's labour on potatoes and bananas, as well as his wife's labour. He also remarked on the loss of his wife's management abilities ('she was the treasurer – she always advised on questions of income and expenditure'). The cropping pattern has changed. They have ceased to grow groundnuts, beans (through lack of labour) and no longer maintain the bananas as well as they might if they had more labour. In addition, Patrick now also has to do the domestic work in addition to the farming. This has meant cutting down on the time spent on farm work. Patrick's relatives do not live in the village, thus they cannot help him – and, he says, local people do not help non-kin.

This household has been markedly affected by deaths – one of which was definitely an AIDS death. The result has been a decreased range of crops, reduced labour inputs to the farm, withdrawal of a child from school (because of lack of cash) and loss of managerial skills and domestic labour. The son contributes some cash from his fishing, but is likely to leave home shortly, thus removing a source of cash income. The foreseeable outcome is that this man will be left alone with his young daughter. It is unlikely that he will be able to remarry in this village, given that he is a stranger and that there has been an AIDS death in the household. It is also worth noting that he has received, and expects, little help from his extended family who live some distance away.

Statistical analysis of household responses

To explore the issues of how people cope with changes in the labour available to households, we analysed the results of a random sample of 129 households. Our respondents were asked to provide details about departures and arrivals in their households and family over the past five years, as well as information concerning age, gender and other circumstantial details. So data on this topic included 'normal' departures (e.g. the marriage of children, the deaths of elderly grandparents) and arrivals (.e.g births, the inclusion of a son's wife in the household). However, the questions also elicited information about AIDS deaths and the movement of children and other family members consequent upon one or more AIDS deaths.

The producer/consumer ratio was calculated for the time of the survey and for five years previously, using a standard FAO (Food and Agriculture Organisation) calibration of consumer and producer units for children and the elderly. Members of the family remitting money, or contributing work or produce, were calibrated as a fraction of a producer, based upon the monetary value of their contribution. The main problem in identifying the demographic impact of AIDS upon production and consumption decisions in these households was the small size of the sub-sample of afflicted and affected households. Our sample was drawn from three main areas and only part of one area (the south-eastern Rakai District) had experienced any marked degree of demographic impact from the pandemic. So the effects of all sorts of other 'normal' demographic changes, often with similar impacts upon the consumption and production of a household to that of AIDS, could not be adequately distinguished from that of the effects of AIDS. Of a total of 129 household responses in our sample survey (including those in the worst hit settlements of Kiembe, Guanda and Kooki, see page 90), eight of these were *affected* by AIDS, 8 households were *afflicted* by AIDS and four households were both affected and afflicted (for definitions of affected and afflicted see page 86). Although the cell sizes are large enough to make some qualitative statements regarding typical responses, it should be noted that they are generally too small to draw statistically significant conclusions.

Figure 6.2 shows the changes in producer-consumer ratios (P/C ratio) over the past five years for all 129 responding households. It can be seen that the frequency curve is skewed to the left (most households have moved in the direction of a more favourable P/C ratio) but that there is a long tail consisting of households which have experienced a wide range of unfavourable movements, some very severe. The majority of these households are AIDS affected and AIDS afflicted. Of the 36 households which were afflicted, affected or both, only two showed a more favourable net movement in the P/C ratio in the last five years.

The detailed responses of households to AIDS was then elicited by means of an open-ended questionnaire. Interviewees were allowed to list any number of responses. So the 129 households gave a total of 167

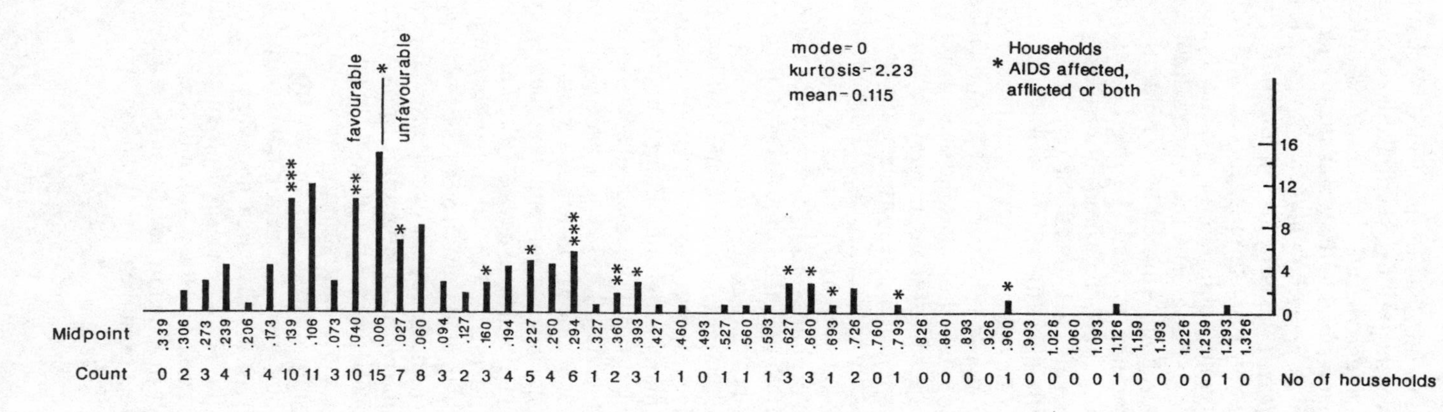

Figure 6.2 Histogram of changes in production/consumption ratios over the last five years

responses to how they had been affected by changes in the producer/consumer ratio over the past five years. Table 6.4 tabulates these responses by the categories *unaffected*, *affected*, and *afflicted*. However, although the cell size is usually too small to draw statistically significant conclusions, most of the differences between *unaffected* households, on the one hand, and *afflicted* or *affected* households on the other, are marked and in the expected direction. For example, affected and afflicted households have a higher rate of taking children out of school than do unaffected, because they are forced to pay greatly increased school fees for orphans who have joined the household. Affected households, too, are most aware that they have more mouths to feed. Hiring labour only remains within the capabilities of unaffected households; only one wealthy household, which had been affected and afflicted by AIDS, was able to afford the hiring of labour. Although afflicted households appear to reduce the area of land cultivated for obvious reasons of labour shortage, the percentage of households doing so is no different from unaffected households. Thus it is not possible to conclude that the overall cultivated area in this region is declining as a result of the demographic impact of AIDS. Simply, there have been many other disruptive events in recent Ugandan history that have also discouraged cultivation. Loss of remittances and other cash income is another important impact of AIDS upon households who have both received orphans and lost household members. This has also been found in AIDS affected and AIDS afflicted households in Malawi (Norse, 1991, 11).

From these data, we can conclude that by 1989–90, AIDS had not yet drawn adaptive responses in production and consumption on a scale that dwarfed the many other adaptations households make all the time in response to other rapid processes of socioeconomic change. However, we believe that in certain localized areas, AIDS is beginning to be *the* major determinant of socioeconomic change and that, in the light of available epidemiological evidence, this situation may be expected to spread to other areas in the region within five years. In certain sub-counties of Rakai District where the pandemic is well advanced and large numbers of people have died, it is likely that adaptive measures by affected and afflicted households have been on a wide enough scale to affect the pattern of agricultural production. Similarly, the demographic impact of the AIDS pandemic upon individual households is difficult to distinguish from the normal domestic cycle unless a much larger sample from different parts of the country is taken. Our main sample was drawn from south-east Rakai, northern Rakai, together with a control sub-sample from Kigesi in Western Uganda; only in the first area had the pandemic clearly affected many households. So it is safe to say that smaller samples drawn from areas particularly affected by the pandemic (see Figure 2.7) would undoubtedly show a higher proportion of responses to a decline in the P/C ratio than we would expect from the normal random movement of the ratio in a stable population.

Table 6.4 Responses to AIDS by unaffected/affected/afflicted categories: percentages by column totals of response

Consequences	Unaffected	Affected	Afflicted	Both	Total
Children out of school	4 3.1%	2 18.2%	1 5.9%		7 4.2%
Less labour on farm	32 24.4%	2 18.2%	4 23.5%	2 25.0%	40 24.0%
More work	15 11.5%	2 18.2%	2 11.8%		19 11.4%
More to feed	16 12.2%	3 27.3%	1 5.9%	1 12.5%	21 12.6%
Hire labour	12 9.2%			1 12.5%	13 7.8%
More labour available	1 0.8%				1 0.6%
More capital used	3 2.3%				3 1.8%
Less land cultivated	17 13.0%	1 9.1%	2 11.8%	1 12.5%	21 12.6%
Change cropping pattern	7 5.3%		4 23.45%		11 6.6%
Less household labour available	4 3.1%		2 11.8%	1 12.5%	7 4.2%
'Lose heart'	4 3.1%				4 2.4%
Form new household	1 0.8%				1 0.6%
Less work needed	4 3.1%				4 2.4%
Yields reduced	2 1.5%		1 5.9%		3 1.8%
Output reduced	2 1.5%				2 1.2%
Cultivate more land	1 0.8%				1 0.6%
Lose remittances or other cash income				2 25.0%	2 1.2%
Increase in domestic work		1 9.1%			1 0.6%
Change to non-agric. activity	1 0.8%				1 0.6%
Increase sale of farm produce	1 0.8%				1 0.6%
All other responses	4 3.1%				4 2.4%
Total	131 100.0%	11 100.0%	17 100.0%	8 100.0%	167 100.0%

Source: Authors' own fieldwork.

Summaries of major household coping mechanisms

The main impacts to be identified from analysis in the preceding sections are:

1 loss of income-earning opportunities in both agricultural and non-agricultural sectors;
2 diversion of productive labour time of still-healthy family members to caring for the sick;
3 diversion of cash to medical expenses, both palliative and in vain search for a cure for afflicted household members;
4 diversion of food reserves to funeral ceremonies and cash for a coffin and other funeral expenses;
5 withdrawal of children from school to reduce cash expenditure and increase available labour time on the farm;
6 altered patterns of consumption and production by households receiving orphans from other households which no longer have adults capable of caring for and looking after children.

These effects, and some typical household coping responses, are presented in Figure 6.3.
Coping responses may also be classified by the *arena* in which they occur, as shown in Table 6.5 below. This table is derived from case material. It does not represent all the logical possibilities.

Coping mechanisms outside the household: the development of supportive relationships – men and women

Taking the case of men first, the appearance of paediatric AIDS cases has led men to support each other in so far as they do not drink alone, which activity they consider may lead to risky sex. While all men will not reform at once, there was said to be a marked decline in visits to prostitutes. Men are scared of AIDS. They visit each other every evening to catch up with news of AIDS afflictions and deaths. Some men still go to brothels and bars, but their behaviour is restrained. Men who have lost several children cultivate in the morning to 'feed the grandchildren' but spend most of the day drinking. Their wives express worry. Men are also taking time to talk to their wives. Most of the conversations concern the need for the family to stick together and be forgiving and tolerant. Some men are confiding their financial and business dealings to their wives – a new practice which began during the Amin period when men were vulnerable to murder.

In the case of women, despite the increased pressure on women's time to cultivate, process food, care for the family and nurse the sick, they are still finding time to listen, comfort and even help each other with chores. Neighbourhood women will appear unannounced to weed and trim the banana gardens of a woman who is ill. They will order one who has been

Table 6.5 Coping strategies in different arenas

1 The homestead	
A Change in household structure:	amalgamation (same generation) splitting additional dependent members (young orphans) additional productive members (older orphans)
B Changes in domestic-work organization:	increase time spent decrease time spent alter work distribution among household members (may affect women)
C Change in level of life/welfare of household members:	poorer diet (restricted range of food, less preparation time) poorer housing (less time for repairs) less access to education (particularly girls)
2 The farm	
A Changes in farm work organization:	increase time spent decrease time spent hire labour
B Changes in farm practice:	decrease cultivated area decrease crop range: cut out cash crops cut out some food crops adopt intercropping
3 The market	
A Changes in cash income:	loss of remittances loss of cash income because of need to use time for domestic/farm work sale of food crops sale of handicrafts other petty trade sale of household labour

Source: Authors' own fieldwork.

nursing a sick relative for a long time to go to the market or visit a friend while they chat with and care for the patient and children. They have persuaded the local Resistance Councils to solicit outside help for the orphans and some have assumed the responsibility of caring for them in their homes. As one woman put it: 'Women know the pains of childbirth. Women know the sadness and loneliness of nursing a terminal patient.

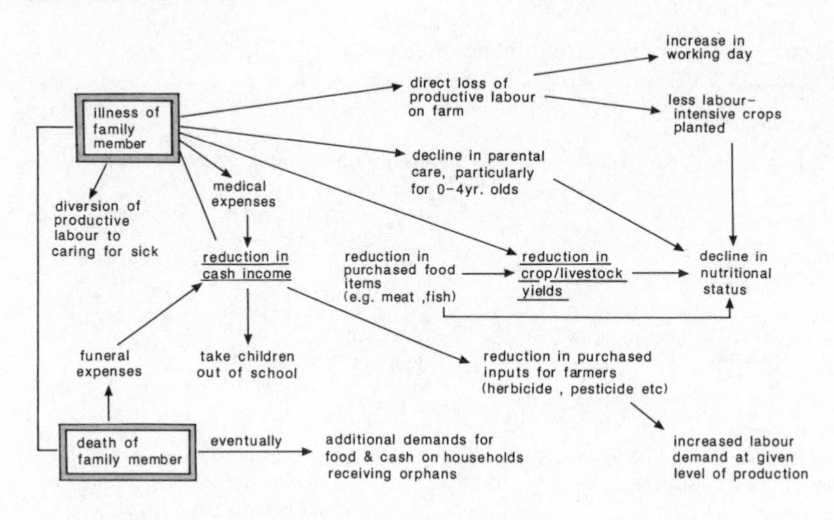

Figure 6.3 Household coping mechanisms in the face of AIDS

Women know that bereavement causes lethargy and yet they know also that the work must be done.' To counter all this, women at the markets, church, roadside and homes seemed to suspend the urgency of other activities and simply listen to each other. Some working women met regularly once or twice a fortnight to relax with each other over a meal or a cup of tea, to talk about their and other women's concerns and to exchange comic tales of AIDS and condoms – relieving stress through humour. The following comments from women in Luanda village may summarize the concerns of all women in these communities. They said:

We all live in daily fear of the new disease.
Women fear what their husbands may bring home.
Women are innocent. They are dying for nothing.
I will not allow my daughter to be married unless the man is AIDS tested.
I wish they would invent a drug for women's protection.

Such informal groups and impromptu meetings are new and they use different spaces. Traditionally women meet up with each other early in the morning or in the evenings while fetching water. These informal counselling sessions are a means for women 'to keep sane' as one woman put it; discussions in Masaka elicited the strongly held view that women require spaces where they can meet to share their common experiences and concerns. There was a feeling that most public spaces belong to men and that their own ability to cope would be enhanced by a clearer demarcation of some space for women.

In Luanda, a predominantly Catholic community, Sunday visiting between households and attendance at public celebrations are considered good for everyone by both men and women. People put considerable energy into helping neighbours prepare for their children's weddings, church celebrations of saints' days and the ordination of priests. These occasions provided increased opportunities for sociability, access to information, exchange of labour and food sharing. One man said: 'Talking to others about death makes it less frightening.' This is why funerals are such important community events. However, these are changing their form as a result of the increasing demands they make upon people's time.

Changes in funeral practices

Changes are taking place in funeral practices. Between June 1989 and March 1990, there were 73 funerals in Lwanda village, a community of around one thousand people. During the busy planting and harvesting months of July–September and February–March, people were overwhelmed by the demands of agricultural work and by the neighbourly demands to spend three non-working bereavement days mourning if someone had died in the village. It seems that the practice now is to keep the bereaved company on the night of the death or the arrival of the body and help with the grave digging, food preparation and funeral arrangements. Thus, grudgingly, but by tacit consensus, the one-and-a-half-day mourning period has become the norm. Previously, funeral ceremonies might last in excess of a week. As one man put it: 'We have to change otherwise if we do not grow food we would be showing irresponsibility to the living'. In *Bukedde*, a Rakai District weekly newsletter, one writer observed: 'Death had become so common that if one misses work for one day to attend a funeral, one can easily end up losing a month's work' (*Bukedde*, 1989, p. 2). It is not uncommon to hear people asking how they are going to manage 'all these funerals'. The following sad but telling incidents underscore the tensions people feel over the loss of production time, through having to take care of the dying and bury the dead.

In Kitembo it was the height of the weeding season and people were anxious to have the work done before the maize, beans and groundnuts flowered, otherwise the crop harvest would be poor. Some people raised a few eyebrows when they started hoeing early in the morning or late at night by moonlight.

In another village, there were four people who had been critically ill for months. One day, two of the patients died. At the funeral of one man whose wife was also found to be critically ill, money was collected for her funeral arrangements. This was considered to be in doubtful taste, but the local gossip had convinced people that she would die at about the same time as the other two, in a matter of hours, and that they would be unable to contribute to and attend all three funerals. In the event, she died 27 days later and the neighbours came to the funeral.

Coping beyond the local community: the role of local non-governmental organizations

Other community-coping mechanisms have also been triggered by the realization that the situation is desperate. Women in villages with many orphans approached a few active members of the local chapter of Uganda Women's Efforts to Save Orphans (UWESO) and, together with the Resistance Council members, arranged for the care of these children. The UWESO chapter has to raise its funds and run income-generating projects for widows so that they can have money to support themselves and above all to send their children to school. Through UWESO's efforts, 20 children have found international sponsors and some blankets and food have been sent to the orphans. UWESO members, RC members and chiefs say that 'things are going to get worse soon'. This has prompted the National Resistance Council member for the worst hit county, Kakuuto, to propose the formation of an Orphans Community Based Organization (OCBO), a Rakai-based NGO that will coordinate the support and relief assistance programmes of the international NGOs and will register all orphans. Identified members of local communities will be trained and made responsible for the welfare of the orphans and their accountability will be monitored by the village Resistance Councils. OCBO was a response to the government's declared desire to reach the orphans through a community-based system of care.

Several NGOs are already involved at the community level with AIDS patients, orphans and other survivors. The Medical Sisters of Mary at St Joseph's hospital at Kitovu have now added an orphan component to their four-year-old programme which provides a home-care clinic for people with AIDS. They have also expanded their activities from Masaka to Rakai District, where they provide medicine, food, school fees and emotional support.

TASO (The AIDS Support Organization founded in 1986) has gained national recognition as its activities have expanded from a few Kampala activists grappling with nursing people with AIDS to the provision of counselling and practical support and training for people with AIDS. Action AID, a British, non-government grassroots organization, supported their efforts from the beginning. Its pre- and post-testing counselling emphasizes positive living based on safe sex and good diet. TASO now has sponsorship from the German Doctors' Foundation and USAID. They have expanded their activities to Tororo town in the east and to Masaka and Mbarara in the west. Both the Federation of Uganda Employers and the Experiment in International Living have launched programmes for AIDS education in the urban work places, but the main focus of their work remains Kampala. Their financial backing comes from USAID and they use the model of trained peer educators.

1987 marked the national and international recognition of AIDS as a national crisis. Uganda became the first country to set up an AIDS Control Programme primarily to monitor and run an education information campaign on AIDS. The Red Cross, the Protestant and Catholic and Muslim Medical Foundations, the churches, UNICEF and AMREF joined in the educational effort which consisted of sermons, booklets, posters and messages as well as radio and TV messages to schools and communities. These efforts are reflected in the awareness apparent among secondary-school children as shown by their essays which were discussed in Chapter 3.

Conclusion

The main conclusions to be drawn from this chapter are that people in Rakai District are currently coping with the impact of AIDS by building upon and adapting existing, well-tried strategies. However, in some cases it is already possible to see that these established methods of coping are coming under pressure; experimentation in terms of funeral practices and the organization of orphanages, to take but two examples, is beginning to take place. It is likely that as AIDS-related deaths increase in the future, more households and communities will find that existing mechanisms are inadequate and further experimentation will take place. Both government and NGOs are beginning to provide additional support. It will be of the utmost importance that the coping capacity of households and communities is closely monitored in order that policies and programmes are designed which can support them as their present coping mechanisms cease to be adequate. It will also be important that these policies and programmes do not replace but build upon the experience of local people.

Notes

[1] Susan Hunter, personal communication, reports that the county clerk in Rakai told her there had not been any marriages in the area in the two years prior to 1989.

[2] This data was derived from a series of extended interviews with households in Rakai District. These households were a random sample of all households in the research area.

7

The Special Case of Orphans

The last chapter explored some of the coping strategies and mechanisms that people in Buganda are adopting in response to the losses of productive labour occurring in the home and on the farm. In both Buganda and the rest of rural Uganda, these two spheres are closely interlinked as they are in all rural subsistence communities. This chapter and the following one look more closely at two aspects of the lives of individuals and communities in which the impact of AIDS is likely to be most important – the care of orphans and the adaptations of farming systems to the loss of productive labour. The common feature linking these domains is labour and the ways in which people cope with changes in its availability. In each domain we see how loss of labour is placing, and will increasingly place, strain upon existing methods of coping at the household and community level and on the farming system as a whole.

In Uganda, an orphan is defined as a child under 18 years of age who has lost one or both parents. Hunter's data collected in 1989 suggested that there were 24,524 orphans in Rakai District. This was 12.6 per cent of all the children under 15 in the district. Of these, 48 per cent had lost only their fathers, 12 per cent only their mothers, while 30 per cent had lost both parents (Hunter, 1989a). No details were available for ten per cent of these orphans. These data did not indicate the cause of parental death, but, given the sudden rise in the rate of orphaning in this area after 1983 and the different pattern of orphaning in, for example the Luwero area (where there were widespread massacres of civilian populations, and the peak of orphaning occurred during 1982–3 (Hunter, 1990), it is safe to assume that many of these cases are in the main the result of the AIDS epidemic. Remember that it is in the nature of the disease that for most orphans the loss of one parent will shortly be followed by that of the other. Here we examine the various fates of these orphans and provide an introductory consideration of some of the social policy implications of this level of orphaning. We describe what happens to orphans in general, to orphans left with one parent, orphans left with their grandparents to care for them and orphans who are left alone without any support from relatives.

There are various ways in which families and households cope with orphan care and the resulting levels at which the children's needs are met are also varied. However, some patterns can be identified in this variation; the most significant explanatory variable is, of course, whether one or both parents have died. However, the different levels of wealth found among households receiving orphans is also important. In considering these issues, there is one important misconception to be put aside. This is the assumption that Buganda is a 'traditional society' in which 'the extended family' will always pick up the pieces when disaster strikes. First of all, as we saw in Chapter 5, Buganda has undergone enormous changes over the past century. Secondly, even in societies where custom is, or was, supposed to dominate, people did not always obey it. Rather, they made their individual and household decisions *in relation to* custom, rather than always *in accordance with* it, as in any human community. Thus what to do and how to do it always contained an important element of individual choice. The importance of this observation is that in Buganda, in Uganda and in Africa as a whole, 'custom' does not dominate to the exclusion of individual, household and familial choice. Responses to abnormal circumstances will undoubtedly increase the extent to which people act in ways which are inconsistent with their customs and values.

Within the Ganda family system, people identify themselves as belonging to large groups described as 'clans'. These clans acknowledge common descent in the male line. The practice of exchanging children between households which are related through networks of clanship has been widespread. The dominant principle guiding these exchanges has usually been that children belong to and are the responsibility of their fathers' clan. Ganda clans are not geographically localized, so children may in the normal course of events expect to spend periods of their childhood living away from their parents and their place of birth. These flexible arrangements offer a range of possibilities for the care of orphans. In practice, children, orphans or not, might expect to be cared for by their mother's sister, mother's brother, father's sister, father's brother, or by their paternal or maternal grandparents at some stage in their childhood. In the case of orphans of Christian parents, there is the option of going to god-parents. More rarely today, there is also the possibility of going to a blood parent where there is a traditional blood pact between their father and another man. Orphans may also be looked after by non-relatives, and both poor and rich families have taken in orphans. The latter seem to have done so to a greater extent, not surprisingly given their greater resources.

Upon the insistence of the dying parent, orphans may also be taken in by friends of the family. Another alternative to have emerged in the Rakai crisis is care by step-mothers. This is seen by the local community as being very unusual: and its singularity is indicative of the extremes to which people are now being driven. In such cases, although these women have severed relationships with the deceased man, either through desertion or divorce, they may now reappear on his death to look after his children from a later relationship.

The enormity of the orphans' problem can be appreciated when we consider that while all of the above caring mechanisms are in operation, some orphans are still left to cope on their own or remain dependent upon other help from outside the kinship system. In the affected communities there is a feeling that more ought to be done and there are many explanations of what is perceived as a reluctance to take in orphans. These centre on the constraints placed on households by the difficulties of the general economic climate. People feel that they can barely feed, clothe and send their own children to school, let alone assume additional responsibility for orphans. Furthermore, given the high levels of AIDS-related mortality in this area, salary earners or the wealthy find themselves increasingly asked to care for the children of their dying relatives or friends. Such people may feel that they are unable to provide for any more dependents in their households. In these circumstances, it may once again be concluded that indigenous mechanisms for managing orphans within the kinship and household system have now reached, or are reaching, their absorptive capacity. There is an additional reason why orphans may remain uncared for. This is the fear that they may be infected with AIDS. Although few instances of this concern were reported, there were sufficient for it to be seen as a significant area requiring public education.

In the next section, we examine the different types of arrangement for orphan care. We distinguish these arrangements in terms of whether children have been singly or doubly orphaned and also of the destination of these children on the deaths of their parents.

Orphans with one parent dead

In Rakai District it was observed that in most cases the father dies first. His death is preceded by processes of change and adjustment within the household as his condition worsens. First of all a woman diverts time that would otherwise be spent in parenting and caring for children to looking after her husband. Then, on his death, she has to continue to divert attention from child-care to agricultural and other income-generating work in order to support herself and her children. Thus, even before the death of a parent, children may have experienced a long period of relative emotional and physical neglect as their mother's energies have been deflected. In order to deal with these problems of domestic and income-generating work, it is inevitable that some children, particularly girls, are taken out of school. In such circumstances, the mother usually stays on in the household.

However, in a minority of cases, women desert their sick husbands. As we saw in Chapter 6, desertion is not an uncommon occurrence in Ganda marriages. It was certainly recognized as a fact of life by the majority of our informants. Because children belong to their father's lineage, when a woman deserts her husband she will often be expected to relinquish her older children to be looked after by her husband's mother or sisters. Very small children who are still breast-feeding will remain with her. If her destination is her parents' home, she may take all her children with her. If she tries to cope on her own, she generally takes only the youngest. Women who desert a sick husband or the homestead after the husband's death go to a roadside location where they can survive through petty trade, to Kampala, to sell beer at a fishing village or trading centre, or to another village where they can find work as agricultural labourers. The general economic insecurity of women leads many to seek patronage and support from one or more men. Urban, trading-centre or roadside locations, with the likelihood of an increased number of sexual partners resulting from economic insecurity, clearly increase the risk of infection if a woman is not already infected, or may serve to spread the infection further if she has been infected by a deceased partner.

In cases where the husband survives his wife, children usually stay on in the parental household and are looked after by their father's mother or sisters. If these relatives live nearby they visit the house to cook, wash clothes, perhaps do some agricultural tasks and provide other care for the orphans. A widower seeks help with the children from his mother and sisters until he is able to remarry. In a typical case, a man who had seven children left the two oldest with his mother, sent three others to stay with one sister, and the two youngest to another sister who also had one of her own babies at breast as well as a toddler. Such an arrangement may be temporary or permanent depending on how the children and the adoptive parents adjust to each other. The children may return to their father and live with him alone or with his new partner, but they may also return to their aunts' homes if he remarries and they do not get on with his new wife. There were a few cases where men remained alone with their infant children and their sisters visited now and then to give advice and some small practical help. Such arrangements are new departures for Ganda men who would not normally play such a role – once again, an indication of the degree to which the abnormality of the situation is causing quite radical adaptations to occur.

The majority of widows stay in the homes where they have buried their husbands. There is relatively little disruption in their social life or that of their children. If relatives live nearby they may provide help with the socialization of children or with labour on the farm. There has been some speculation that in-laws might try to drive widows away from their husband's property and this has occurred in some cases. In other cases, widows who do not get on with their in-laws have nevertheless stayed on in the village. It is important to remember here that fathers and sons do not always live on the same or adjacent land, so any conflict can be minimized by social avoidance. As long as the widow and her children have their own house, they can ignore minor conflicts; in any case, the children are soon old enough to help with the agricultural labour and thus in time in-laws become less important for the survival of this type of female-headed household.

Some widows with one or two young children (probably under ten years of age) do return to their parents' homes or migrate to the urban areas. The parents are always supportive of widowed children who thus receive assistance from them with the socialization of children and with labour. Grandparents are usually left with the children if the widow decides to remarry or go to the town. This established practice precedes the advent of AIDS. Some widows with a child or two may leave them with a sister or close friend who has children of about the same age. Children of widows who go to live elsewhere are encouraged to visit their paternal relatives and get to know them. This is important to their later lives and is a marked advantage which they have over children who enter institutional care where they may lose touch with their kin.

Double orphans

When both parents have died there are four destinations for orphans: to stay in the parental home, to go to their grandparents, to go to guardians or to go into some kind of institutional care.

Remaining in the parental home

Children may remain in the parental home if there are no grandparents or father's sisters with whom they can live. This occurs because the grandparents are either dead or too old to look after orphans, and in the case of the father's sisters, if they are unwilling or insufficiently wealthy to assume the extra burden. It is usually the wealthier families who can afford to look after orphans. In cases where the eldest orphan becomes the head of the household, clearly the age of the children and the number of very young siblings is crucial to survival. In these circumstances, relatives and neighbours sometimes provide assistance in the form of cooking or helping with house repairs on the basis of goodwill and individual discretion rather than upon the stipulations of 'custom'. So such arrangements are optional and cannot be relied on. Inevitably, in many cases care may not be available.

Where one parent has already died, it is at the funeral of the second parent that the fate of the orphans is discussed in the light of the verbal or written will of the deceased. In the absence of instructions, relatives will volunteer to take up a certain number of children. Sometimes children may attach themselves to a relative or to a close friend of the family or to an acquaintance.

Throughout rural Rakai, there are homes that have been abandoned, closed up and deserted, where the children have gone to live elsewhere, but everyone in the community knows to whom the land belongs. Even so, it has been suggested by some observers that it is best that children stay in their parents' homes so as to protect their rights in the land. This view expresses the widely held concern that orphans may be deprived of their rights in the land and other property. Local Rural Resistance councillors and village elders can defend an orphan against a father's brother or a landlord who may wish to usurp the children's rights. This is an important issue given the considerable confusion that exists in Buganda in relation to questions of land tenure. However, most orphans stay on in the parental home simply because they have no close relatives who wish to take on the task of rearing them. The cases that follow serve to illustrate the range of variation in level and strategies for coping that characterize these 'orphan households'.

CASE 1

In the first household, there are four children aged 11, eight, six and three. Their father died in 1987 and their mother in 1988. They continue living in the parental home and are looked after by two aunts who are their deceased father's sisters. From time to time these children may also spend time in their paternal grandfather's house as a respite from the pressures of coping alone. The two aunts produce food for the orphans, and their father's two other married sisters living in adjacent villages help with school fees, clothes and additional care and medical expenses if the children are ill. The three oldest children are in school, but have all had their schooling seriously disrupted by their parents' lengthy illness. As a result they all remain in the first class of primary school.

This interconnected-care group based on kinship has no difficulty in providing adequate food for the orphans. The problem comes with paying for things that require money. The family says that it has chosen to care for the orphans in this way to avoid the loss of their home and their land. It is significant that the two women who take primary responsibility are both divorced. Although they have their own children, they are able to provide the support the orphans need, perhaps because they have no husbands.

CASE 2

The second case is rather different because the oldest child is already adult and has her own children. Joy is a 20 year-old single mother of two; for the last five years she has managed to keep her three brothers at home together and even pay their school fees. When the parents died the relatives who lived in the nearby villages encouraged these children to stay in the parental house. Their father had been a travelling salesman and their mother a farmer; both died of AIDS within one year. They left no money but they had built a decent house and had a respectable banana plantation. Relatives were unable to offer these orphans any financial or labour support, so Joy made and sold mats in order to raise money. Her girlfriends helped with this handicraft production. Joy cultivated the farm to provide food for her siblings. She comments on these early years immediately after the deaths of her parents as follows: 'It was hard work. The boys at first felt rebellious but we soon came to terms with the need for work and discipline. Our two paternal and maternal uncles always included us in wedding and other celebrations. This was a comfort. We remain in close touch with our family, and today I am preparing food because we are going to celebrate the ordination of my maternal uncle.'

This particular group of orphans has been so self-reliant that now some relatives seek help from them. They are also engaged in helping two other orphan households nearby. Joy explains why their own experience has led them to do this. She says: 'The greatest help is having people to talk to. We help with the gardening and cooking because one home has only boys who, although they are learning fast, do not really know how to care for themselves, while in the other home the children are still very young.'

Despite her responsibilities, Joy is looking to the future. Although she had to leave school when she was 15 years old, she wants her brothers to go to university. She hopes that selling bananas and making mats will continue to generate money. Her lover, the father of her two children, suddenly married someone else when he found out that she was an AIDS orphan, but she hopes to marry at some time in the future.

CASE 3

The third case shows a group of siblings surviving without relatives but with a lot help from the community. There are five of them, three girls and two boys aged 14, 12, 11, 10 and nine years. They live in an isolated house on the edge of the village. Their parents died of AIDS two years ago. The house remains half completed, without windows, the kitchen has collapsed to the ground and the remains of an old car stand in the area behind the house. The children have a small plot measuring 20 metres by 30 metres on which they grow some tomatoes and onions. These they sell to generate income. For food they maintain a small patch of bananas and they also grow beans and potatoes. As soon as their parents died, they dropped out of school because there was no money. The only relative they know is a maternal uncle who lives in another county. They are fond of him but he is not dependable because he drinks excessively.

The children are rarely at home. They have made themselves indispensable by helping with the funeral rites of their community. The three elder children take turns in sleeping at wakes to keep company with the relatives of the deceased, while during the day they all go to help with the additional household chores associated with the presence of visitors. Neighbours admire their courage, diligence and hard work. They are checked upon by neighbours, receive gifts or food and an agricultural officer has encouraged them to start a small horticultural enterprise to bring in some cash. Their level of life compares favourably with that of other, non-orphan, children of the same age and circumstances in the area. Their growth is not stunted from malnutrition and they do not have to beg for food.

CASE 4

The last example of an orphan household demonstrates a very different response from the preceding cases, showing how remnants of different households may come together to form a new household. It is significant that the members of this household are all women, and shows us something of the ways in which women may support each other in the face of their generalized economic vulnerability.

Twenty year old Anne, a bank employee in Kampala, died of AIDS in January 1989 and was buried at her maternal great-grandmother's home in Rakai. She was survived by her infant son and daughter, her mother Bitulensi, her grandmother Sarah and her great-grandmother Leya. Leya was bequeathed 20 acres by her husband. She raised three children, but only Sarah is still living and she moved in with her mother many years ago upon becoming widowed. Sarah had three daughters, two of whom are married and live far away, but Bitulensi, who deserted her husband 20 years ago, lives with them. Mary, her granddaughter, lives with them and has three children. Anne's two babies have always lived with them.

These women grow enough food on which to survive, but Anne (the deceased) used to send money for extras like meat or sugar and the needs of the children. She also provided them with clothing. When Anne became ill she decided to stay in town where she could more easily obtain medical attention. She only returned home when she was close to death. Her relatives visited her regularly and at great expense. In the end they had to borrow money to cover some of the funeral expenses. Subsequently, Bitulensi and Mary have been hiring themselves out as agricultural labourers to repay the debt. This has caused them to miss the planting season in their own gardens, and means that their family will have fewer beans and groundnuts to eat just when the meat and other extras have become less available as a result of Anne's death.

With households of young orphans, the age and gender of the eldest child are crucial. Even when girls are as young as 12, meals are cooked, water is fetched and some limited cultivation takes place. In contrast, with boys of the same age, concerned village women have often had to intervene to stop them from having to beg for food and to avoid malnutrition. These differences are illustrated by the following two case studies.

CASE 5

Nineteen-year-old Nankya, as a 12-year-old girl, assumed responsibilities for her five siblings aged two to nine. She says: 'We raised ourselves on cassava, beans, potatoes and spinach because that is what was easy to grow and to prepare'. She has managed to keep the four boys in school as well as her young sister. She has also started making mats to pay their school fees. On Saturdays the family works for long hours weeding the banana garden and tilling or planting crops.

Previous examples have shown households where the children are coping reasonably effectively. The last example stands in contrast to these.

CASE 6
This is a household of six physically stunted children who look old for their age, headed by a boy aged 12. Immediately after the deaths of their parents, the neighbours provided them with food whenever they asked for it. Before long, they went around the village begging. A seven-year-old brother was even found seeking work in town. The local headmaster said that the three children, a girl and two boys who attended school, came irregularly and in shabby uniforms. A young widowed neighbour has recently volunteered to provide them with a decent meal a day. This case has been reported to the local councillors and to the local NGO, the Ugandan Women's Efforts to Save Orphans (UWESO), in an effort to save the household from disintegration.

Orphans who go to their father's sister or mother's brother

In Buganda, a father's sister and mother's brother usually play a major role in the parenting of children even when their parents are alive. Children born out of wedlock can be brought up by the mother of one of their parents. This happens in cases where an unmarried man asserts his paternal rights and takes the child away from its mother, or when an unmarried mother finds it convenient to hand over the baby to its paternal home. Similarly, there may be occasions where an unmarried woman refuses to acknowledge the father of her child or the father refuses to acknowledge paternity and the woman passes the child to her mother to be reared. According to custom, the mother's brother disciplines and supports those of his sister's children who have not been officially redeemed by their father. However, under normal circumstances children may spend as much time in their father's sister's and mother's brother's homes as they do in that of their parents. If siblings are on good terms, parenting is easily shared. However, some of the orphans are not being acknowledged or claimed by these relatives and it does not appear that differences in wealth provide an explanation. Individual preferences, personal grudges (and possible fear of AIDS) may be responsible for this avoidance of customary responsibilities.

Orphans with grandparents

Inevitably, many orphans are cared for by their grandparents. While this results in particular problems of discipline and financial resources, it should also focus our attention on the additional and important problems associated with the care of the aged in communities where the 'mature' generation may be severely depleted by AIDS-related deaths during the next ten years. Some observer's (Beer et al., 1988) have suggested that AIDS will inevitably result in a 'grandmothers' burden', where the grandparents take on responsibility for the care of their orphan grandchildren. Evidence for this can already be seen from data in Chapter 6. However, there is also another type of grandmothers' burden: the burden of an old age unsupported by adult children. The situation of such a person is, at present, fraught with difficulties. In the course of our research, we came upon several cases of very elderly people who, for various reasons unassociated with AIDS, had no children to support them in their declining years. Old, ill and weak, they scraped a living by working on their own farms or as labourers on somebody else's. Living in isolated households, they cared for themselves as best they could. In communities where many children will never reach adulthood, a proportion of today's 40 and 50-year-olds must look forward to a bleak future. In communities such as those in Buganda, where mutual aid flows along *effective* kinship lines, the outlook for isolated old people is not good.

However, at present, aged grandparents have increasingly assumed responsibilities for rearing orphans. Sometimes they are the only relatives left. Grandparents typically lack the energy to work long and hard in their gardens, so the range of food available to them and their dependents becomes smaller and household nutritional status falls. Household income is unlikely to be supplemented by cash as the elderly are unable to cultivate crops for the market. The implications of this were summed up by one teacher who said: 'We have over 80 orphans in this school and you can tell those from grandparent homes. Their noses are always running and their uniforms are often not clean, ironed or repaired.' Also, some young people were to be found playing truant in the nearby town and were identified by members of the community as orphans coming from grandparent homes.

Many grandparents with orphans confessed to the problem of discipline. In the words of one: 'I cannot force them to leave for school on time. I never know when the Parent-Teacher Association (PTA) meetings are held.' Attendance at PTAs by grandparents who are carrying the burden of orphan care is important because the households without orphans need to be informed of the reduced economic status of orphan families in order to enable the headmasters to commute the PTA fee. In 1989, many orphans in the research area had dropped out of school because their grandparents could not afford these fees. A teacher and a grandparent independently pointed out that: 'grandparents are often ignorant of how, or lack the energy, to go and defend their wards, or at least enable them to get a fair hearing and treatment in cases involving insubordination to school authorities'.

Most of the orphans in the research area were in households where the grandparents were in their forties and fifties. These people often had younger children of their own and the orphans appeared to adjust to the circumstances very well. Aged grandparents in their fifties and sixties have also assumed a major role in taking care of orphaned grandchildren. However, such grandparents, with their diminished capacity for work, cannot provide adequate material and economic support for the children. These households were vulnerable to malnutrition and infectious diseases because of low food production and because medical care could not be paid for. In addition, in Ganda society there is an expectation that grandparents will be indulgent, passive and permissive towards their grandchildren. This results in role conflicts where the grandparent is expected to fulfil both the parental and grandparental roles at the same time. There was anecdotal evidence that since orphans from grandparent families are not disciplined enough to accept and respect school authority, they were more likely to drop out of school. Where there is a single female grandparent, the situation is exacerbated. Women typically lack money, do not have the same entitlement to community labour as men and depend to a greater extent on friendship-based goodwill.

Guardians

Some orphans are looked after by guardians, either in their old home or in those of their guardians. These guardians are usually neighbours or those appointed by relatives who cannot take the responsibility of looking after the orphans. The experiences of these orphans are mixed. They are very vulnerable to the loss of their land rights. There were a number of cases before the Rural Resistance Council courts where ***mailo*** landowners had claimed the property of the orphans under the 1928 Busulu and Nvujjo Law (West, 1972, pp. 69–94). This accorded and protected the occupation rights of tenants as long as they continued to pay a specified fee; it was a contract between the landlord and the *male* peasant and not with his wife or children.

Orphans without relations

Orphans with no relations to look after them and who cannot remain in their community, are accommodated within various arrangements. These range from a recognized foster-mother who may take in orphans in addition to her own children or 'house parents' encouraged by community organizations to look after orphans in designated homes, to 'orphanages' run by resourceful individuals who undertake this responsibility from a sense of social duty or as a way of earning an income. The Orphans' Community Based Organization (OCBO) has been proposed as a way of consolidating all these efforts, to attract overseas funding and to check the possible abuses inherent in unsupervised informal arrangements. On the whole, the orphan problem is most visible where UWESO members are active, and invisible where members are inactive or the organization does not exist. In badly hit counties like Kyotera, some local councillors and UWESO members have collaborated to find widows with large homes to look after orphans who are too young to live on their own. The case of Safina shows how these arrangements work:

CASE 7
Safina, the 30-year-old widow of a trader who had also been married to three other, now deceased, wives, was in charge of 12 children (six of her own and six from her deceased co-wives). The UWESO members

requested that she use her large home to provide day care and meals for some local children aged between 12 and three who were living by themselves. The Rural Resistance Council provided the food and UWESO tried to provide clothing and toys.

Other examples of this type of community response are where UWESO has acquired an abandoned home and installed an aged couple to look after the orphans under the supervision of a neighbour who was a teacher, and the case of a group of elderly women with their orphaned children who were brought together in the large house of a man who had died leaving very young children and no known relatives. In two cases, individuals have opened up orphanages and only informed the authorities when they were fully operational. These individuals usually have connections with fundamentalist churches in Kampala or with relief agencies like the Red Cross. These private orphanages sometimes give rise to problems. In one home, for example, the couple sold the goods donated to them to support their alcoholism. They had initially received help because in the past they had performed other charitable work. Furthermore, apparently eight of the 12 children were not orphans but belonged to various relatives of the couple. In general, though, such arrangements appear to be providing a satisfactory level of care in small units that mimic the normal Ganda household.

In contrast to such small caring units, there are also larger scale orphanages. For example, there is the case of a local clergyman who acquired land situated near a river covered with reeds which provided a breeding ground for mosquitoes. Here he constructed some inadequate huts. Although this establishment received donations of food, medicine, hoes and seeds, it was grossly overcrowded. Six teachers and 209 orphans were housed in a five-roomed mud-and-wattle building measuring 10 metres by 20 metres. For a long time, this miserable group of children shared a few blankets, cups, plates and basins. Inadequate nutrition, malaria and the physical exhaustion of the children from farm and domestic work were evident. Every Wednesday nurses came to treat any of the children who were sick, but without any accommodation suitable for the segregation of children with infectious conditions, illnesses tended to spread and recur. The older children said that most of the time was spent discussing religion but on Wednesdays, when the nurses came to visit, other subjects were written on the blackboards that were placed under trees for the five different classes.

The main problem with this establishment was that many of the children were not from Rakai District. A careful examination revealed that they had been recruited from as far away as Mbarara in the next district, and some from unmarried mothers who had been persuaded that their children would have an education and a better life in this place. Some of these children seemed to have a recollection of their parents but they were inhibited from speaking of them by the presence of the supervising adults. In addition to the residents, this facility was also open to local orphans who brought their own lunch and attended during the day. This orphanage had received recent donations of bedding, medicine and food from a number of sources. Whatever the motivations of the founder of this place, conditions were clearly unsatisfactory, and it shows the need for some kind of regulation, either through government or through linking it with an NGO, either national or foreign.

The Orphans Community Based Organization may provide such a supervisory facility. This arrangement has been proposed by concerned local people in anticipation of government plans to reach orphans through community-based programmes run by foreign NGOs like Save the Children Fund and World Vision. OCBO might facilitate the smooth running and coordination of the programmes through the registration of orphans, operating income-generating activities and ensuring the general well-being of orphans on a community basis. The proposal suggests among other things that members of the community be identified and trained to care for the orphans and also for people with AIDS, providing these services through the local Resistance Councils. In this way, support, relief and assistance intended for the orphans may have some chance of reaching its target. The community will be made responsible for the welfare of its people and accountability could be monitored using the Resistance Councils. These efforts could then be supplemented by help from organizations such as UNICEF, Save the Children Fund and TASO. However, in Uganda it is inevitable that some people are cynical about the role of the Resistance Councils and other bodies which they see as trying to line their pockets in the name of orphans. So in this as in any locally based initiative, accountability is of the utmost importance because ordinary people are sceptical that aid to orphans may be misappropriated. The problem is one of the lack of good faith against a background of the breakdown of civil order and communal responsibility. In Ugandan circumstances this is clearly something that must be taken into account in any voluntary/NGO activity.

Meanwhile, UWESO continues to help the Resistance Councils to register orphans (particularly in Kyotera, Kakuuto and Kooki counties), and to find school sponsorship for the orphans in schools where the commutation of orphans' school fees is an important issue and from various foreign agencies and individuals working in the Rakai District. UWESO depends upon the voluntary efforts of a few middle-class professional and overworked women. In some places personality clashes and jealousies between Resistance Councils and UWESO members have adversely affected the registration and educational sponsorship of orphans.

The special problems of orphan households

It is now appropriate to attempt some summary of the practical implications of the preceding observations. The case material on orphan households shows that they can be viable. This is to be expected, given that the unviable ones are likely to have disappeared and we have no information about their decline. Certain features seem to be conducive to the survival of those orphan households that do survive more or less intact. In terms of holding the siblings together, it makes a difference if the eldest child is a girl. Girls learn early how to perform household and nurturing chores. The proximity of relatives is also important because it provides psychological and social support for the orphans through their inclusion in family gatherings and celebrations. This ensures that the orphans remain integrated into the local community much as they would be if their parents were still alive. This is important if the relatives are unable or unwilling to take time to offer labour or financial assistance. Indeed, it might be suggested that for the orphan households, involvement in community funeral activities, which thereby builds up a system of reciprocal entitlements, is of considerable importance. Indeed, it might be counselled that the continued integration of these young people into their communities should be a major goal of any policy towards orphans. The prospects of large numbers of orphans (the current number of orphans in Rakai alone, leaving aside the future orphans in this district and also throughout Uganda, is already very great) detached from their communities, alienated from the mainstream of their national life and devoid of land, must be of the greatest concern. In order to confront some of these worst effects of the AIDS epidemic, the following policy areas require urgent attention.

Education

Those dealing directly with orphans and their problems see their education as a priority. The younger children need to start school and the older ones must be kept in school. UWESO members and the District Administrator in Rakai have been lobbying headmasters to commute the orphans' school fees. In fact, though, the school fees are not the main problem. Rather, it is the relatively large fee levied by the Parent Teachers Association to help pay the teachers or pay for school projects in the absence of adequate subventions from government, which presents the major obstacle. Many parents can afford a term's fees of, for example, 180 shillings, but cannot afford the 3,700 shillings per year for the PTA. It is a struggle for parented children to stay in school; this struggle is greater for orphans. The community is, however, divided on this issue. Many parents insist that every child must pay the fee and that commutation for the orphans might create a privileged class of educated orphans.

The content and quality of the education these orphans receive is also important. A case could be made that it should be predominantly agricultural or vocational, for all of these children will have to earn their own livings from an early age. Indeed, some vocational schools run by the religious orders are accepting a limited number of orphans because they believe that the future of orphans will depend upon their having marketable skills, as not all of them will wish to live by agriculture. However, such a view effectively excludes a large number of people from other, more academic, types of education, and in so far as Uganda may lose many of its skilled and trained people to AIDS[1] these will have to be replaced. So it can be argued that the orphans created by the epidemic should not be doubly penalized by exclusion from the best education available.

Parenting

While these responsibilities are taken on in some cases by a father's sisters and younger grandparents, there has to be some doubt whether the full demands of caring for, teaching, socializing and disciplining children can any longer be met by the extended family. As the case material illustrates, the coping capacity of some households and kin networks is already being exceeded. Many influential people in Uganda say that the orphaned children should be absorbed into the extended families because it is 'traditional'. This assertion is prompted by the opposition to orphanages that are assumed to take children out of the local community. These arguments ignore the fact that most of the AIDS orphans will probably grow up with non-relatives anyway and that there is an urgent need to establish institutional care which will be community-based but run by professionals in education, nutritional health and child-care. This may also relieve the burden on the grandparents who experience problems in providing adequate care. What is clear is that decisions now need to be taken, and policies developed, to deal with this problem of how best to achieve care within and in relation to the community without producing a generation of children who are institutionalized and alienated from their society.

Nutrition, shelter and clothing

We have seen that in the great majority of cases these fundamental needs are met. In so far as rapid appraisal can in any way be a reliable guide, there were few orphans, except at the unsatisfactory orphanage which has been described, who showed obvious signs of failing to have these needs satisfied. However, there is a noticeable minority of orphans who showed signs of malnutrition, with the under five-year-olds exhibiting the extended stomachs and bleached skin and hair associated with malnutrition. Some of the older children also appeared underfed and ragged. These children were not neglected: they were simply receiving inadequate nutrition, particularly those who were living in poor households. Their condition was a result of being orphans in a poor society, rather than of simply being orphans. In some cases there was not enough food, particularly during the dry season. The special nutritional and care problems of these children need to be given special attention. But it is also important to note that in subsistence communities affected by AIDS, the amount of labour available for food production will decline. This problem may become more pronounced in the coming decades, not only in Uganda but also elsewhere in sub-Saharan Africa. This more general issue is examined in detail in Chapter 8.

Legal protection of orphans' property

In the course of the research for this book, we heard of a number of disputes before local courts involving orphans. These cases concerned the rights of orphans to inherit their deceased parents' tenure of the farm. It would be a worrying precedent if landlords won any of these cases. Clearly, these children are vulnerable to pressures from unscrupulous guardians, relatives and others. Minors are always vulnerable to relatives who may try to cheat them out of their inheritance, or who may divert their wealth to educating their own children. Orphans will be no exception. In Rakai a few landlords had started dispossessing the orphans of the rights in land which their parents had enjoyed and which are widely treated as heritable in these communities. The local people supported the orphans in their claims at court. However, it should be noted that the threat to the orphans reflects wider changes in the society and economy of Uganda. This threat is predominantly from the Kampala-based *nouveaux riches* Rakai expatriates who made money from smuggling during the 1970s and 1980s. These people are now using their accumulated capital to buy up large tracts of land to start commercial agriculture or ranching. In some cases, they offer to buy the land; in others they press tenants to leave through the threat of legal action. In one instance the District Administrator engaged in a prolonged defence of an eviction case on behalf of the widows and orphans. No legal decision had been reached by the end of the research period, and while the widows and orphans remained on their land, their security was not guaranteed. This may indicate more widespread insecurity, and thus an additional burden for households already under stress through death and illness.

In all this, there are specific problems of policy that need to be tackled and tackled soon. However, the underlying policy issue is the choice between targeted and non-targeted aid. The former may create privileged households and encourage corruption and even orphan-farming by some unscrupulous people; the latter may be seen at best as an inefficient use of scarce resources, at worst as a threat to the welfare of these children. In general we are talking about very poor people and very poor communities, their poverty made worse by the past traumas they have suffered and the current trauma of AIDS. The choice of how to pass between the Scylla of targeting and the Charybdis of non-targeting must surely be one of the cruellest this society could have to face. It is in the care of orphans that it faces it in an acute form.

Note

[1] On some projections, it seems that in a country where the AIDS epidemic has reached the magnitude it has in Uganda, in order to have one person aged 50 surviving from the current cohort receiving education and training, 17 will have to be trained now. Personal communication from Jane Rowley.

A Note on further reading

In addition to Hunter's work, most accessible in Hunter, 1990b and Dunn, Hunter, Nabongo, Ssekiwanuka, 1991, other material on the orphan question includes: Preble, 1990, and Mukoyogo and Williams, 1991.

8

The Impact of AIDS on Farming Systems

Africa is the only region of the world where overall food production has declined in recent decades. In most sub-Saharan countries the average index of food production per head shows a marked decline between 1971 and 1981 and 1986 and 1989 (United States Department of Agriculture, 1990). Per capita consumption in 1980 was 15 per cent lower than at the start of the 1970s and a full 20 per cent below that at the start of the 1960s. This suggests that even in normal times a large proportion of the population of Africa has a seriously inadequate diet and per capita calorie intake falls short of acknowledged minimum nutritional standards. African economies are essentially based on rural production. Even the more industrialized countries such as Kenya and Zimbabwe have dominant agricultural sectors, responsible for some 80 per cent of GDP. So the ability of the small producer to self-provision is already under threat and the AIDS pandemic may constitute an additional but decisive shock to some systems of agricultural production. The rural labour force employed in commercial farms and plantations may also suffer significant losses. This has additional, serious implications for countries that are heavily dependent on the export of primary agricultural products, such as Zimbabwe, Kenya and Malawi.

One response by governments to this secular decline in food production over the last two decades has been to increase imports of staple foods. This strategy, along with other problems such as market imperfections, centrally determined pricing policies weighted against the producer and a decaying physical infrastructure, has often had the effect of further weakening the position of the small producer. Although there were some indications of improved food output in certain areas of the continent in the period 1985–6, recurrent drought in the Sahel and in Central Africa, together with severe market imperfections, suggests that the long term decline is likely to continue.

Against this background, the potential impact of AIDS over the short, medium and long terms may be extremely serious. African rural production systems are often inherently fragile, particularly in the drier regions, and, as we have seen, the destructive consequences of AIDS are potentially enormous. It is important that efforts are made to monitor the impact of the pandemic on food supplies with a view to identifying those areas that are most vulnerable to labour losses. Such an exercise should have a specific policy goal.

We have argued that AIDS is a long wave disaster. If this is so, then the aim of rational policy should be to develop responses for the agricultural sector which are ahead of the worst impact of the pandemic. Such responses will obviously include labour-substituting agricultural technologies, but, perhaps not so obviously, also labour substituting *domestic* technologies (such as piped water) which release additional time to women whose role in African farming is so vital (Boserup, 1970) and whose role in coping with the effects of AIDS is, as we have seen, so great.

With these issues in mind, this chapter examines the present and possible future impact of AIDS on rural livelihoods in Uganda, while considering some more hypothetical scenarios in other areas of Africa. While Chapters 4 and 6 discussed the impact of the disease and the ways in which individuals, households and communities coped with the epidemic, here focus is directed at the farming system level. The underlying assumption is that AIDS will reduce labour availability on the farm, eliciting a number of coping strategies. Farming systems are therefore analysed according to their patterns of labour use, including the seasonal peaks demanded by crops that are currently grown as well as those that could feasibly be substituted within ecological constraints. The typical arrangements for organizing labour within the household according to gender and age, and the implications of altered labour allocations because of AIDS mortality for food outputs, need to be considered. Upon these considerations, farming systems in Uganda are classified according to their potential vulnerability to labour loss. Then the growth of the AIDS pandemic through space and time is overlaid upon a map of farming systems of differing vulnerability, and the currently observed and future hypothetical impacts are analysed.

Households and farming systems

The household livelihood model in Chapter 4 identifies livelihoods, which are combinations of income opportunities, each of which has certain qualifications of labour, capital rights and other resources. Households have a varied range of these qualifications and earn livelihoods of varying security and quality. Most of the income opportunities in rural Uganda are agricultural: specifically crop and livestock combinations. These vary geographically, according to broad opportunities and constraints laid down by interacting variables of climate, altitude, soils and vegetation. They not only vary with these physical data but also with other, broader socioeconomic data. Thus for example, the same physical data can pertain to two adjacent areas: one a communal area of smallholding farmers and the other a large commercial farm. Clearly the income opportunities for households in each of these areas are very different. One conceptual tool for classifying and analysing these agronomic and socioeconomic differences is the farming system.

A farming system is an abstraction. It can be seen as a number of components linked by the flows involved in agricultural activities. Typically the components are human beings, soil, water, crops, livestock, implements, trees and so on. Flows include energy in different forms, including labour power, nutrients, fodder, food for humans, compost, crop residues and cash. These flows may be quantified in terms of energy flows, quantities of various material inputs and outputs or in monetary values. Analyses may be made of the relationships between any number of components. For example, the effects of the introduction of improved fodder for cattle can be traced through to an increase in draft power, more timely ploughing and higher crop yields. The importance of socioeconomic variables such as the pattern of labour allocation, the types of agricultural technology, the cultural preferences in production and consumption and the relative prices of the outputs and inputs into the system are very important, although it may be fair to say that the subject matter of farming-systems research is more agronomic and economic than anthropological (a point which is taken up below). Also, system variables can be analysed and descriptions of system properties such as 'resilience' and 'sensitivity' may be used to describe the response of the whole system to external changes or shocks – in this case labour loss as a result of the impact of AIDS.

Farming Systems Research (FSR) has been a major research and policy tool for at least 20 years (Collinson, 1982). It is as well to rehearse briefly the purpose of this approach before using some of its results for the different purpose of tracing the impact of AIDS. It was developed in order to put individual technical improvements (e.g. crop breeding and improved agricultural implements) into a comprehensive framework. The piecemeal development of single technological innovations was frequently met with baffling (and baffled) lack of enthusiasm by the farmers of Africa, Asia and Latin America. Too often, the technical development of a single component in the system had unforeseen effects on others. Farmers often did not innovate because they had to bear the indirect costs of the innovation in another part of the system. Thus, additional labour demands on households, customary food preferences and local systems of land tenure were sometimes totally ignored by the research station that had developed the 'improvement'. In response to such problems, a wider ranging analysis was developed in which changes in one part of the system could adequately be traced through to others that had been ignored by the terms of reference followed by plant breeders who were working on research stations far from the world of farmers.

Farming systems are also used to classify farming over geographical space. There are usually problems of boundary definition and it is often necessary to introduce intermediate or interdigitating zones where different system variables are changing at different rates over geographical space. There is also the problem that the farming system has to generalize from a range of different farms and recourse is usually made to ideal-type farms or mean values for farms within the system. Thus inter- and intra-household variations in resources, wealth and food security have to be subsumed in the farming-system approach. In other words, the farming system defines the 'full permissable range' of income opportunities, but consideration of the patterns of household access are also required in order to determine which household can choose what production and consumption strategies from this range. So in using the farming system as the level of analysis for understanding the impact of AIDS, it is impossible to predict how a certain type of household within each farming system will feel the impact without more detailed information being provided by household or farm management surveys. In virtually all attempts to use farming systems data to predict the impact of AIDS, recourse has to be made to a range of representative farming households within each system. It should also be recognized that FSR is largely based upon economic and agronomic criteria and ignores insights drawn from anthropology altogether. The latter are important because they address the cultural aspects of adaptation and the ways in which people make decisions about productive adaptation to the impact of AIDS on criteria other than purely economic ones. They are able to draw upon resources from a much wider frame of reference than that implied by FSR.

Two notable studies of the possible future impact of AIDS were commissioned by the Food and Agriculture Organization (FAO) and the British Overseas Development Administration in 1988 and 1989. These were able to use farming-systems data and detailed farm-management studies collected for other purposes some years before to simulate the impact of AIDS on farming communities in Rwanda and Tanzania. Gillespie (1989a, 1989b) used a large, detailed farm-management survey in Rwanda, dividing the country into five farming systems on the basis of altitude, soil fertility and population density. These primary system characteristics, together with additional data from farm-management records, determined the seasonality of labour demand, the degree of labour specialization, the interdependence of labour inputs, the labour economies of scale and the ecological potential for the substitution of less labour-intensive crops. For each system three ideal-type households with different profiles of labour availability were identified. Then one, two and three deaths from AIDS in each type of household in each farming system were hypothesized. The calorific value of each crop, together with the maximum monthly labour demand, were calculated. From this exercise, the relative vulnerability of the five systems was ranked. The most vulnerable system was that in the Volcanic Highlands of Rwanda because of the marked seasonality of labour demand and the sexual division of labour. It was found that within each system, small nuclear households were more sensitive to labour loss than larger, extended ones.

The Overseas Development Administration commissioned a similar study for Tabora District, Tanzania (ODNRI, 1989). From existing farm-management studies, food and cash requirements, labour availability, land availability and requirements for food production were computed. On to these data were superimposed different levels of HIV infection and mortality from AIDS. Three ideal family types were specified (small, nuclear; larger, nuclear; and extended - with 12 children and eight adults), each in three different agro-ecological zones (high rainfall with low fertility soils; low rainfall with medium fertility soils; and low rainfall with high fertility soils). Sequences of one and two deaths were run on these six family type/agro-ecological zone combinations. The worst effects of AIDS were experienced by nuclear families, especially the small ones, in all zones, and those families of all types in the second zone because of a combination of poor soils, lack of a viable cash crop and the maldistribution of access to land.

Both these studies are most useful in combining quantitative data at the household and farming-system level. They are good early guides to the possible impacts of AIDS for policy-makers and indicate the types of households and farming systems that may have AIDS-induced food crises in the future. However, the multitude of coping mechanisms identified in Chapter 6 indicates that people's resourcefulness in the face of a disaster such as AIDS has probably been underestimated. Without empirical data on how households actually cope in the circumstances, it is difficult to know what the outcome at the farm-system level will be. Governments, and most of the planning information on which they rely, tend to overestimate the resources of poor rural producers but to underestimate their resourcefulness. Planners therefore should be aware that there still remains quite a high degree of uncertainty about what the impacts at the farming system level will be. Ground checking of such simulations wherever possible is, therefore, to be recommended.

There is, though, another set of factors that were not considered by these two studies. These revolve around compensatory flows of various kinds in response to differential shortages of labour and land reflected both in the market and outside. Some of these complex knock-on effects are discussed in Chapter 9; only one, that concerning migration, is examined here. Migratory labour either to urban areas or to other agricultural areas, petty commodity production and trading are important alternative income opportunities and will feature in any coping mechanism. These also operate at the farming-system level. For example, in areas where there is exceptional pressure on land and consequent out-migration, labour losses may be made good by male members of a depleted family returning to claim and work the land.

The degree to which a farming system is vulnerable to the AIDS pandemic depends upon a number of characteristics; most of these are linked to adaptations to the loss of labour. How they are defined is frequently determined more by the availability of data and the form in which these are collected. For the purposes of this analysis of farming systems in Uganda, the classification of 50 farming systems by Johnson and Ssekitoleko (1989) was used but the quality of data leaves much to be desired. Thus, as the authors note:

Uganda has no information whatsoever on its agricultural land resources . . . all quantitative information given in this report should be regarded as a rough estimate, although it is probably closer to the mark than anything else available. (Johnson and Ssekitoleko, 1989, p. 7)

Bearing this and the limited range of useful data in mind, the following primary characteristics to measure the vulnerability of labour loss through AIDS were chosen:

1 Whether the farming system already has a shortage in energy or protein. While it may be possible for individual producers to adapt successfully to labour loss in a variety of ways, the existence of systemic food shortage in the first place suggests a lack of access to land, labour, or capital, making adaptations more difficult.
2 Whether labour supply has less than an excess of 20 per cent of present labour demand at any time during the agricultural calendar. The product-mixes of different farming systems vary in their labour-demand profile. In general terms the more seasonally peaked the demand, the more vulnerable the system will be to labour loss. The 20 per cent point was selected to represent the loss of between one and two people from an average farm on account of AIDS-related deaths.
3 Whether there are substitutable staple crops requiring a lower level of maximum labour input, which will provide sufficient energy and protein. This characteristic assumes that there will be adaptations in the direction of less labour-demanding crops within ecological constraints.

This latter part of the algorithm which searches for alternative crops to make up energy or protein shortfalls within existing labour constraints requires an additional search algorithm for alternative crops. A successful search must fulfil three criteria:

a) The crop must be grown by at least half the farmers.
b) It must yield the increased energy and/or protein to the level required.
c) The net labour required for the shift in cropping pattern must not exceed 80 per cent of the existing labour supply from the average farm household.

For example, farming system 7B in Table 8.1 has a shortfall of five labour days in the month of peak seasonal demand. This could be overcome by growing an increased area of bananas, but the first-year labour demand to establish banana gardens cannot be met from existing labour supplies. Short-term cassava, however, is another alternative, but again there is not enough labour. Thus the search for alternative crops in this case was unsuccessful.

The algorithm used to classify farming systems into a ranking system of vulnerability to labour loss is shown in diagrammatic form in Figure 8.1. It can be seen that there are four categories of vulnerability that may

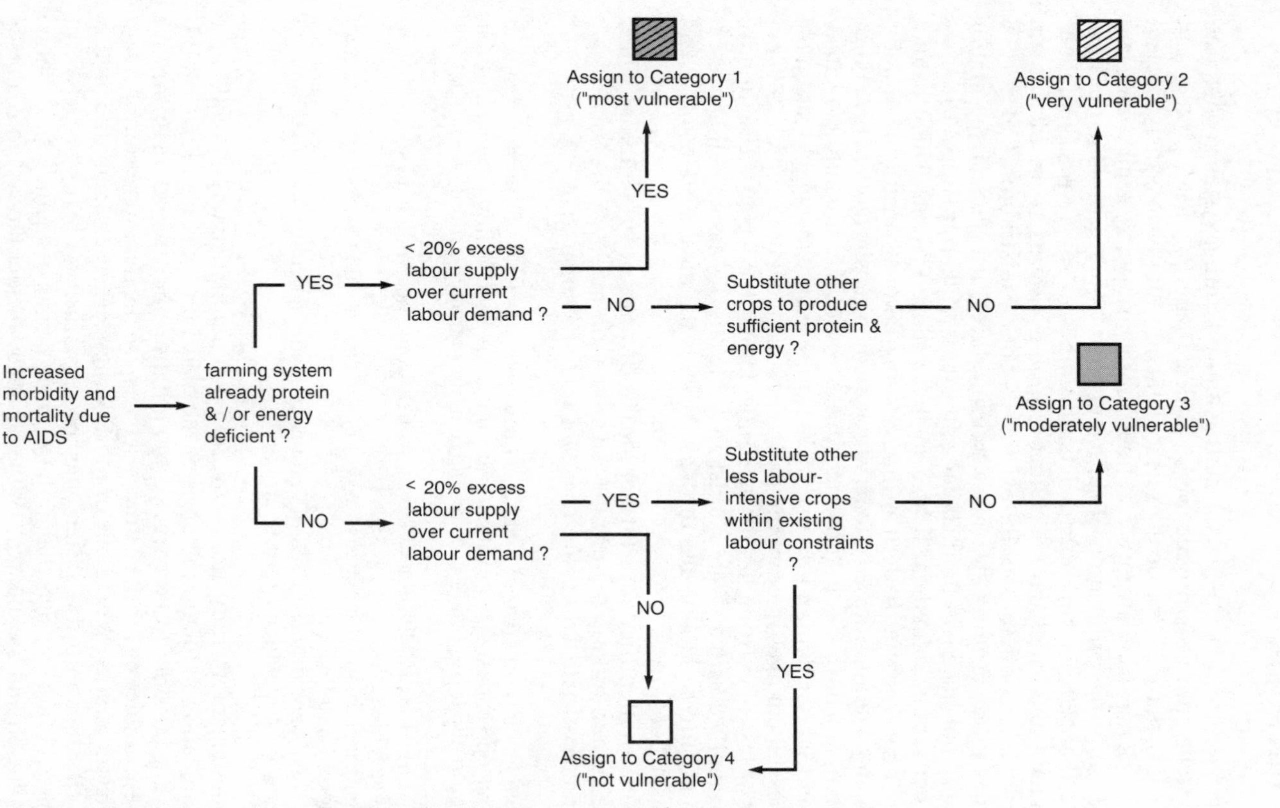

Figure 8.1 Algorithm for classifying farming systems according to their vulnerability to labour loss as a result of AIDS

be derived from it. Category 1 defines those farming systems which are both deficient in the production of sufficient energy and/or protein and have less than 20 per cent excess of labour supply to satisfy present labour demands. Category 2 defines those farming systems which are deficient in the production of sufficient energy and/or protein, but have more than 20 per cent excess labour supply to satisfy present labour demands, and which cannot change cropping patterns to provide enough energy and protein within existing labour constraints. Category 3 defines those farming systems which produce sufficient energy and/or protein, but have less than 20 per cent excess of labour supply to satisfy present labour demands. It also includes those systems which cannot alter their cropping pattern within existing labour constraints. Finally, Category 4 defines those systems which are neither short of energy and/or protein, nor have less than 20 per cent excess of labour supply to satisfy present labour demand. It also includes those systems which can make successful changes in cropping patterns to reduce labour inputs to 80 per cent or less of existing labour demands. Those that were successful have been marked with a single asterisk in Table 8.1. It is assumed that present vulnerability to labour loss through AIDS is the most in Category 1 and least in Category 4.

We also identified four further local indicators of vulnerability. These were used to modify the primary classification made on the basis of the characteristics listed above. These local characteristics are:

1 Significant economies of scale in returns to labour involved in major agricultural operations. (Two examples are those forms of shifting cultivation that require the transportation of vegetable matter for the burning and the herding of cattle);
2 A marked sexual division of labour, which would accentuate the loss of labour for specific tasks;
3 Significant essential maintenance cost of soil conservation works, such as terracing and bunding;
4 A significant degree of out-migration and non-agricultural income brought about by population pressure on land.

These last four indicators were used in a qualitative way to modify the classification made on the basis of the algorithm shown in Figure 8.1. It was not possible to construct a quantitative and precise algorithm for these variables as for the primary set because the data were simply not of sufficient quality. Where one or more of these criteria was used to modify the classification, the system has been marked with two asterisks in Table 8.1. Table 8.1 lists the main farming systems of Uganda as identified by Johnson and Ssekitoleko (1989) and notes whether the system is already energy and/or protein deficient; the present maximum-labour supply in an average household minus present peak seasonal demand; the length of effective rainy seasons in days; the total rainfall; and the crops grown by 50 per cent or more of farmers. Figure 8.2 shows the results of this analysis in cartographic form.

Table 8.1 The main farming systems of Uganda and their vulnerability to labour loss

Farming systems	E & P Energy/protein deficient? yes/no	Max supply minus peak demand for labour	Length of effective rainy seasons (in days)	Total rainfall (in mm)	Crops grown by > 50% of farmers	Comments & Classification
1A	E No P No	8	192	1335	sweet potatoes 63% maize for grain 53%	4
1B	E No P No	0	160	1294	Finger millet 58% Sweet potatoes 70% Maize 66%	Vulnerable on labour criteria. Some increase on finger millet possible. 4*
1C	E No P No	5	211	1513	Cassava 70% Finger millet 58% Cotton 53% Sweet potatoes 52% Maize for cobs 55%	4
1D	E No P No	1	212	1422	Cassava 68% Sweet potatoes 58% Maize for grain 70% Finger millet 76% Cotton 66% Cowpeas 53%	Vulnerable on labour criteria, but adjustments possible to make good labour shortfalls. 4*
1E	E Yes P Yes	4	173	1211	Bananas 82% Robusta 69% Sweet potatoes 60%	Lightly populated. Tse-tse infestation, much forest. 1 Vulnerable on both protein & energy criteria. Some possibility for finger millet.

Table 8.1 Continued

Farming systems	E & P Energy/protein deficient? yes/no	Max supply minus peak demand for labour	Length of effective rainy seasons (in days)	Total rainfall (in mm)	Crops grown by > 50% of farmers	Comments & Classification	
2A	E No P No	−15	174	1140	Bananas 90% Cassava 56% Arabica 65% Cowpeas 81%	Vulnerable on labour criteria. Major coffee growing area.	3
2B	E No P No	0	161	1456	Bananas 90% Cassava 56% Arabica 80%	Vulnerable on labour criteria.	2
2C	E No P No	12	182	1281	Bananas 90% Cassava 56% Arabica 71% Maize for grain 96% Field beans 83%		4
3A	E No P No	−7	181	1242	Cassava 86% Sweet potatoes 77% Cotton 67% Finger millet 69% Groundnuts 55% Sorghum 58%	Vulnerable on labour criteria. Some switching to cassava possible, but not sufficient labour.	3
3B	E No P No	−10	215	1309	Cassava 98% Sweet potatoes 83% Finger millet 95% Sorghum	Vulnerable on labour criteria. Limited possibilities to switch to sorghum, but not sufficient labour.	3
3C	E No P No	−7	197	1123	Cassava 85% Finger millet 68%	Vulnerable on labour criteria.	3

4A	E Yes P Yes	−5	77	656	No crops grown by > 50% of farmers	Vulnerable on protein/energy & labour criteria. Very dry pastoral area, low population density. Cassava would provide energy, but not possible because labour constraint.	1
4B	E Yes P Yes	−3	85	850	No crops grown by > 50% of farmers	Vulnerable on protein/energy & labour criteria. Very dry pastoral area, low population density. Cassava would provide energy but not possible because labour constraints	1
4C	E No P No	−18	86	724	No crops grown by > 50%	Vulnerable on labour criteria. Pastoral system, very small agricultural as very small population	3
5A	E No P No	5	231	1376	Cassava 84% Pigeon peas 74% Maize for grain 56% Field beans 55% Finger millet 73% Cotton 68%		4
5B	E No P Yes/No	−25	194	1325	Cassava 89% Sweet potatoes 90% Maize for grain 79% Field beans 66% Finger millet 81% Sesame 60% Pigeon peas 73%	Vulnerable on labour criteria, particularly because of peaks in July/August. Some crop diversification possible.	3

Table 8.1 Continued

Farming systems	E & P Energy/protein deficient? yes/no	Max supply minus peak demand for labour	Length of effective rainy seasons (in days)	Total rainfall (in mm)	Crops grown by > 50% of farmers	Comments & Classification
5C	E No P No	−5	183	1245	Cassava 70% Sorghum 67% Sweet potatoes 78% Cotton 94% Maize for cobs 54% Pigeon peas 59% Finger millet 65% Groundnuts 66%	Vulnerable on labour criteria. Some switching to maize possible. 3
5D	E No P No	10	186	1381	Cassava 70% Sorghum 55% Sweet potatoes 79% Cotton 68% Maize grain 76% Cowpeas 68% Finger millet 78% Groundnuts 68%	4
6A	E No P No	−14	194	1311	Cassava 60% Maize grain 60%	Vulnerable on labour criteria. some switching to maize possible. 3
6B	E No P No	12	153	1013	Cassava 60% Finger millet 55%	4

6C	E P	No No	−17	162	1174	Cassava 60% Maize cobs 54%	Vulnerable on labour criteria. Some switching to sorghum possible.	3
6D	E P	No No	22	190	1443	Cassava 60%		4
6E	E P	Yes? No	−5	168	1073	Cassava 60% Maize cobs 69% Finger millet 50%	Vulnerable on labour criteria.	1
7A	E P	No No	11	207	1309	Bananas 90% Field peas 56% Cassava 65% Finger millet 54% Sweet potatoes 100% Maize cobs 61%		4
7B	E P	Yes/No No	−5	200	1429	Bananas 90% Cassava 66% Sweet potatoes 74% Maize cobs 63%	Vulnerable on labour criteria	1
7C	E P	No No	9	82	761	Bananas 90% Cassava 66% Sweet potatoes 71% Maize cobs 63%		4

Table 8.1 Continued

Farming systems	E & P Energy/protein deficient? yes/no	Max supply minus peak demand for labour	Length of effective rainy seasons (in days)	Total rainfall (in mm)	Crops grown by > 50% of farmers	Comments & Classification
7D	E No P No	−1	179	1048	Bananas 90% Cassava 66% Sweet potatoes 64% Maize cobs 52%	Vulnerable on labour criteria. 3
7E	E No P No	−9	163	1245	Bananas 90% Cassava 66% Sweet potatoes 79% Maize grain 85% Field beans 50%	Vulnerable on labour criteria. 3
7F	E No P Yes/No	−1	234	1499	Bananas 90% Cassava 66% Maize cobs 57%	2
7G	E No P No	−22	191	1760	Bananas 90% Cassava 66%	Very vulnerable on labour criteria. Mountainous/ high altitude 3
7H	E No P No	4	110	1021	Bananas 90% Cassava 66% Field beans 62%	4
7I	E No P No	6	132	1009	Bananas 90% Cassava 66%	4

8A	E No P No	6	159	963	Bananas 98%? Sweet potatoes 58% Field beans 50%		4
8B	E Yes/No P No	1	157	1424	Bananas 98% Sweet potatoes 53% Field beans 56% Finger millet 59%	Vulnerable on both labour & energy criteria. Well watered & some substitution with other crops possible.	1
8C	E No P No	14	108	865	Bananas 98%? Field beans 71%	Traditional grazing area.	4
8D	E No P No	9	155	970	Bananas 98%? Sweet potatoes 53% Field beans 50%		4
8E	E No P No	4	155	970	Bananas 98%? Field beans 50%		4
9A	E No P No	8	174	1204	Bananas 92% Maize cobs 80%		4
9B	E No P No	6	202	1337	Bananas 92% Finger millet 61% Sweet potatoes 69% Sorghum 75% Maize grain 80% Field beans 58%		4

Table 8.1 Continued

Farming systems	E & P Energy/protein deficient? yes/no	Max supply minus peak demand for labour	Length of effective rainy seasons (in days)	Total rainfall (in mm)	Crops grown by > 50% of farmers	Comments & Classification
9C	E No P No	−5	163	997	Bananas 92% Sweet potatoes 68% Maize cobs 83% Field beans 59%	Vulnerable on labour criteria 3
9D	E No P No	1	74	710	Bananas 92%	Vulnerable on labour criteria 3
10A	E No P No	4	131	1021	Bananas 82% Robusta 69% Sweet potatoes 94%	4
10B	E No P No	11	142	1228	Bananas 82% Sweet potatoes 59%	4
10C	E No P No	−3	88	773	Bananas 82% Robusta 69%	Vulnerable on labour criteria. Limited possibility, if substituting short term cassava. Cattle area. 3**
10D	E No P No	1	190	1201	Bananas 82% Robusta 69%	Vulnerable on labour criteria 3 Marginal for bananas. Some possibility for substituting cassava.

10E	E No P No	2	190	1386	Beer bananas 82% Robusta 69%		4
10F	E Yes P Yes	<1	158	2111	Bananas 82% Robusta 57% Sweet potatoes 79%	Vulnerable on both labour & energy/protein criteria. Suspect data. Most of population involved in fishing.	1**
11A	E No P Yes/No	−3	200	1297	Bananas 95%	Vulnerable on both labour & protein criteria. Marginal for bananas, fairly dry, some cattle-keeping. Few available alternative crops.	1
11B	E No P No	5	186	1240	Bananas 84% Robusta 89% Sweet potatoes 61% Maize cobs 50%		4

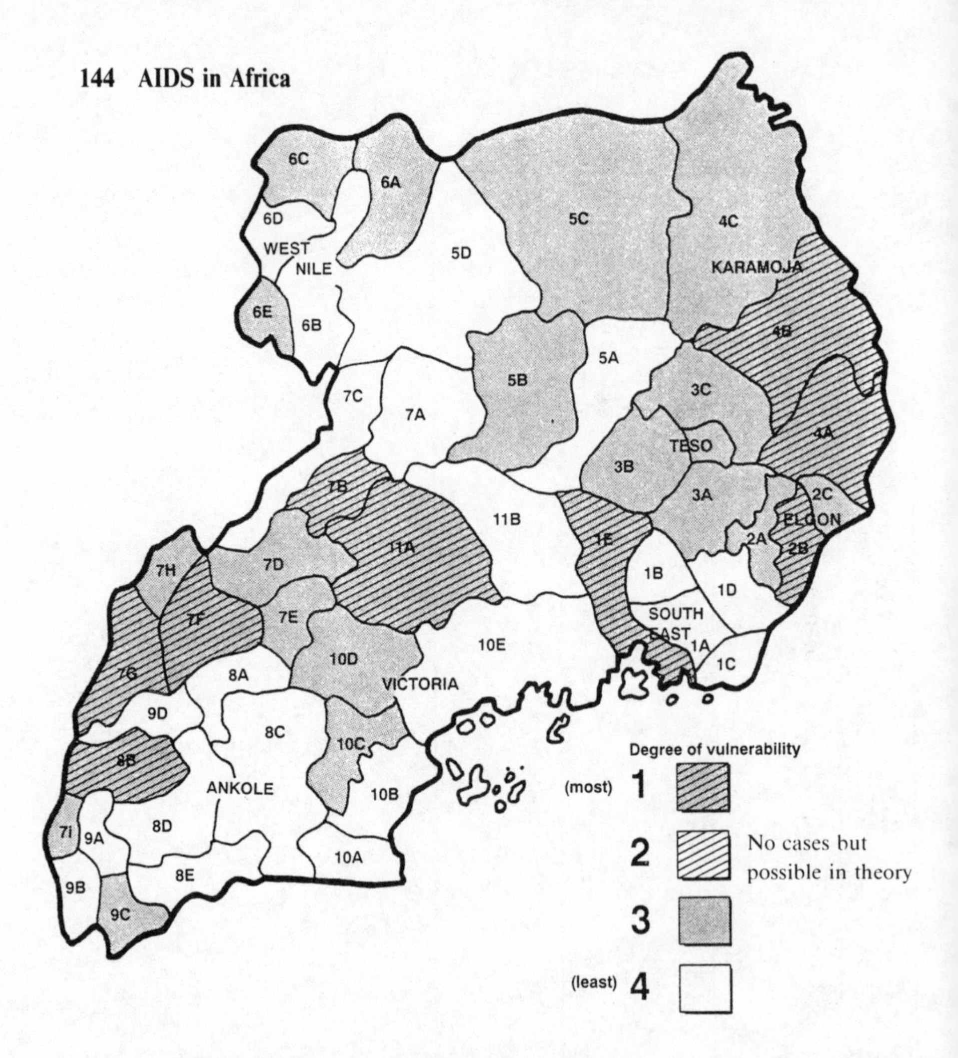

Figure 8.2 Vulnerability of farming systems to labour losses as a result of AIDS in Uganda

Source: Authors' calculations, based on Johnson and Ssekitoleko (1989) and other data, as discussed in text.

The next task was to map levels of seropositivity and superimpose these on the map of farming-system vulnerability. There are data for seropositivity at the district level from the national serosurvey carried out between September 1987 and January 1988, but these are seldom made available to researchers. However, data at the regional level were available and these were mapped. The information is at a much coarser level than that concerning farming systems, but was the only data set available. Also, the data are at least four years out of date, a fact with which all planning agencies have to contend. It is likely that the actual numbers of seropositive people is in the order of one and a half million, and not eight hundred thousand as indicated in the 1988 National Serosurvey. Also there will have been spatial spread from the centres of earlier infection.

A map of endemic areas drawn from 1985 data (Okware, 1987, not reproduced here) shows the areas of infection being highest along the international highway joining Malamba on the Kenyan border, through Kampala and along the northern shore of Lake Victoria, towards the border with Kagera District, Tanzania. 'Newly infected areas' are marked as Masaka District, further west towards Mbarara and the far west around Kabarole and Kasese. Thus it is possible to put a little more geographical detail on the regional-level data. However, there are also data on the distribution of AIDS cases on a district-wise basis provided in the Quarterly Surveillance Reports from the Aids Control Programme in Entebbe. While this provides more spatial detail, it is not a reliable guide to future levels of mortality, because of selective reporting in different areas of the country (for example, the civil war in the north of the country will undoubtedly have reduced reporting there). The population of each District (from the National Census of 1980) was then divided by the cumulative deaths from AIDS by district of residence to give a rate of deaths per head of population. Thus there are two data sets - one based on a reputedly fairly reliable serosurvey, but very outdated and available at the regional level only; the other on district-level information, up to date but suffering from undoubted incompleteness and selective reporting. An eclectic method of estimating district-wise seroprevalence was adopted using a variety of sources (Okware, 1987; Carvalho et al., 1989) which, together with other district-wise serodata which were made available informally, enabled the construction of a fairly reliable account of the present levels and distribution of seroprevalence in Uganda. District-wise data from the National Serosurvey are in the possession of the Uganda AIDS Control Programme but these were not available to us. (Figure 8.3 shows the cumulative rates of AIDS mortality and has been used with other evidence to categorize districts into four levels of seroprevalence (see Figure 8.4).

Both these maps support Okware's earlier evidence, but the more recent data on AIDS mortality indicates that the disease is well-entrenched in the north of the country. The conclusion we must draw is that we simply do not know in any detail either the spatial configuration of mortality or seropositivity, although we know more about the latter but with outdated accuracy. By superimposing Figure 8.4 (rates of seroprevalence) upon Figure 8.2 (potential vulnerability of farming systems to labour loss), it is possible to identify those systems at present under threat. Those systems most affected are:

Under serious threat:
11A (in northern Mubende and western Luweru Districts)
7B (northern Hoima District)
7F (Kabarole District)
7G (Kasese District)

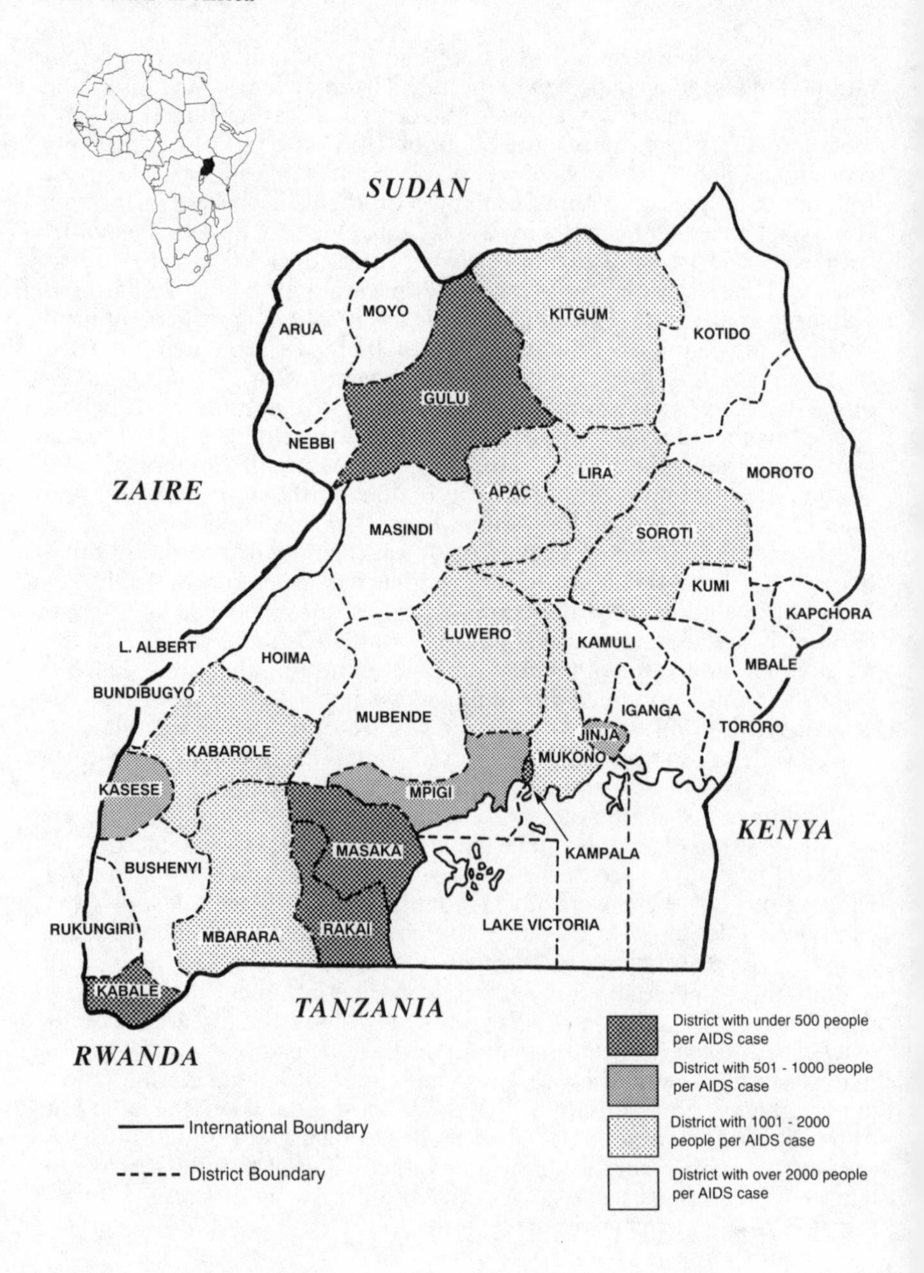

Figure 8.3 Cumulative rates of reported AIDS mortality, Uganda

Sources: Uganda National Census, 1980; AIDS Control Programme, Entebbe, Uganda.

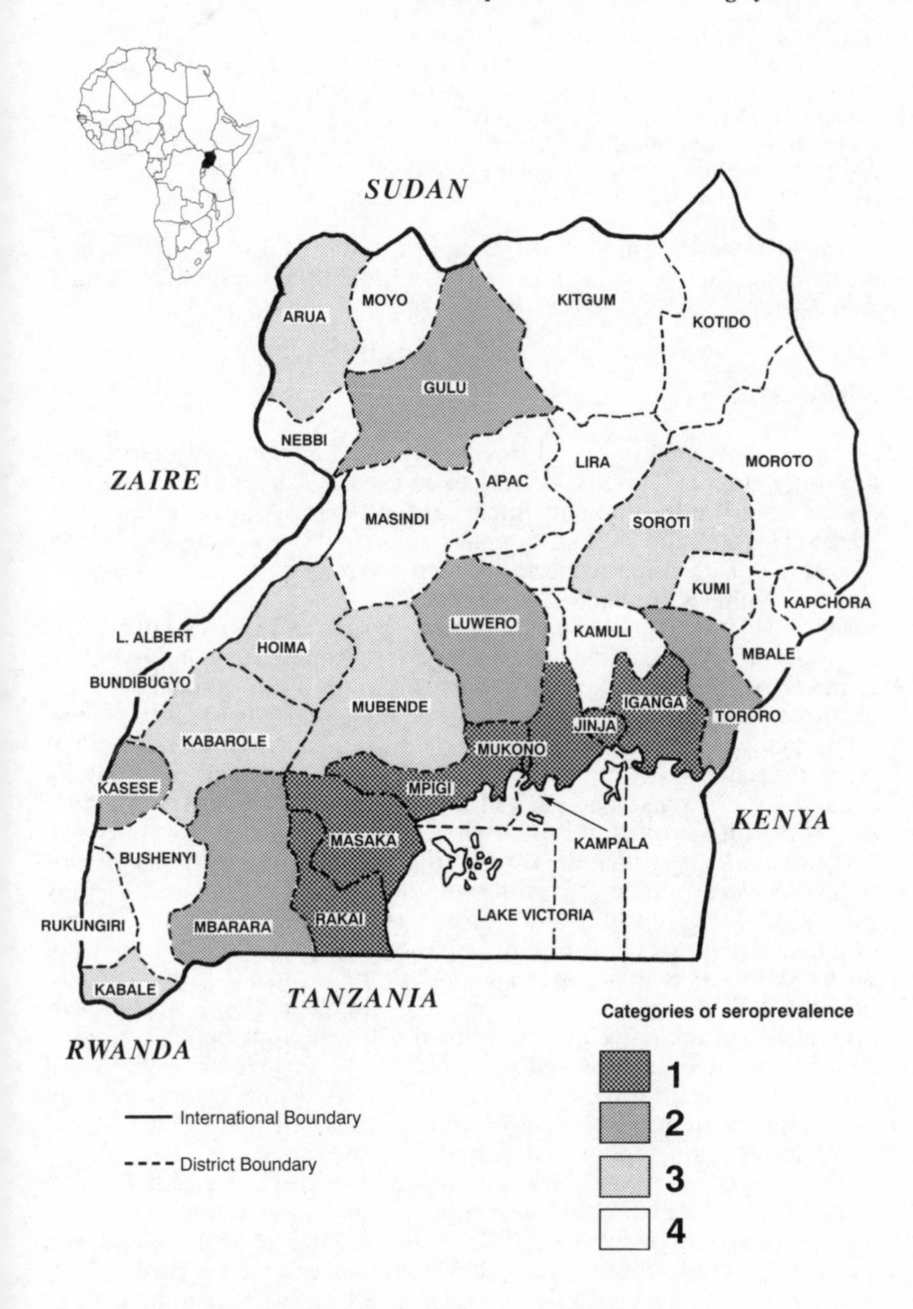

Figure 8.4 Estimated district-wise seroprevalence, Uganda

Sources: Various. See text.

1E (southern Iganga District)
2B (Tororo District)

Under moderate threat:
10D (south-western Mubende District)
10C (north-western Masaka District)
7D (southern Hoima District)

It is also worth noting that the highest levels of seroprevalence and mortality happen to be in those areas with the least vulnerable farming systems.

Ground Checking

Fieldwork was undertaken in three locations, each situated in different farming systems. The household survey used in Chapters 6 and 7 was drawn from two locations in Rakai District, corresponding to Farming System Nos 10A and 8C, and from Kigesi District in Farming System No. 9B. Additional data were collected from interviews with agricultural officers and from other secondary data sources.

The first system studied was a banana, coffee and two-season annual crop-farming system of the Lake Victoria foreshore in Rakai and Masaka Districts (No. 10A). This system is characterized by a low seasonal variation in the demand for labour. The labour demands for perennial tree crops are virtually constant throughout the year (McMaster, 1962). The annual crops (cassava, sweet potato, yam, Irish potato, groundnut and others) usually have a two-season peak of labour demand. There is also a high degree of crop choice and it is possible to retreat to less labour-intensive crops incrementally. Perennial tree crops are relatively forgiving of temporary neglect, particularly if the ground surface is mulched, thereby inhibiting weed growth. It is not surprising therefore that in the 20 years of upheaval and social disruption experienced by Uganda, there has not been famine or even widespread food shortages in this area. Even in households of young orphans, it has been possible to grow enough to survive, although the range and nutritional value of crops may be reduced. This capability of incremental retreat and wide range of choice is one of the reasons why, in Chapter 6, it was difficult to identify clear patterns of agricultural response, particularly crop-switching, in spite of the fact that exceptional demographic change had been experienced.

The second system studied was a cattle and small-stock pastoral system with minor banana and annual crop cultivation (including cassava) in north-west Rakai District (No. 10C) and also found to the north and west in Ankole. While the remarks about banana cultivation made above apply here, too, they have limited relevance because only a minority of households has access to lands suitable for banana cultivation; even for these, cattle-keeping is by far the most important activity. Important economies of scale in returns to labour can be found in cattle-herding, particularly for larger herds and in the dry season where the cattle and herders have to travel large distances. There is also a sexual division of labour in the herding, milking and care of cattle. A reduction in the availability of adult labour will have a much more severe impact upon the system, causing overgrazing near water sources and homesteads and shortages of specific skills and fulfilment of roles in the cattle economy.

None the less, there is a minor seasonal peak of labour demand in cattle-keeping here, because of increased labour requirements to take cattle to more distant watering places during the dry season. Banana as the main crop of the area is, as has been indicated, a crop without marked peaks of seasonal labour demand. Thus, in spite of a number of subsidiary factors that might have made this system vulnerable, labour supply exceeds demand (see also Kyeyune-Ssentongo, 1985).

The third system, a sorghum, sweet potato/Irish potato and bean-farming system is situated in the high plateau (1500-1800m) of Kigesi District (No. 9B). This system occupies moderate and steep slopes and thus requires investment in and maintenance of soil-conservation works. It is an area characterized by extreme land pressure and long-established seasonal and more permanent out-migration of males. Seasonality of labour demands is not marked and losses of labour to a high degree can continue to be absorbed by returning migrant labour. There is also a fairly wide choice of crops with labour requirements that differ in seasonal peak times. Therefore we concluded that this system will be resilient to sudden demographic change. These findings for the three farming systems studied are supported by the secondary data analysed from Johnson and Ssekitoleko above.

Possible implications of AIDS for other farming systems in Africa

The level of sophistication of the classification of farming systems in Uganda developed in the previous section is probably appropriate, given the limited data available and the known complexities of household-coping strategies. Those studies that utilized data from farm-management surveys will be able to predict the growing impact of AIDS upon production with more accuracy, but will still leave a great deal of uncertainty.

In spite of these difficulties, it is perhaps useful to attempt to develop a few simple hypotheses concerning the degree of vulnerability of other farming systems to labour losses. These may take the first steps towards robust generalizations about the effects of AIDS upon farming in Africa and the possibility of mapping farming systems in relation to their sensitivity to labour loss across the continent. The starting point is the degree of seasonality of labour demand in terms of hours of work per head per day. A major determinant of this is the length of the wet season (and, correlated to this, the quantity of rainfall). Figure 8.5 (from Richards, 1952) shows the spatial-temporal distribution of wet and dry seasons in the tropics.

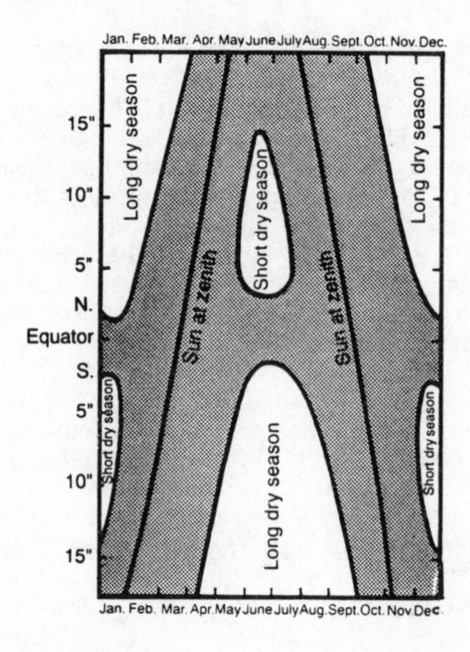

Figure 8.5 Wet and dry seasons in the tropics

Source: Richards (1952) p. 139.

Within about 17 degrees north to 17 degrees south, the rains have the most marked effect on the seasonal pattern of employment (Cleave, 1974). The main impact is the longer the rains, the greater the flexibility the farmer has to stagger farming operations through time, the less severe the penalties for delaying planting, weeding and harvest times for particular crops, and the greater the range of crops that is ecologically feasible. For example, Cleave (1974, p. 128) draws attention to two cases only 130 miles apart in Sokoto, Nigeria, where the southernmost villages have a month's longer rainy season and also slightly more rain. This gives the farmers there the flexibility to avoid a clash between the labour demands for cotton and food grains, which is denied to the farmers further north. The late planting of cotton in the former villages leads to lower than optimum yields, but the farmer can grow more cotton at a lower opportunity cost and total yield value for the whole farming system is maximized. While the onset of the rainy season(s) usually marks the beginning of the period of the most intense work, this is not invariably the case where the opening of new land and the cutting and transportation of vegetative matter in preparation for burning, for example, can be carried out in the dry season. None the less, the generalization linking the length of the wet season to severity of labour demand is a robust one, although it could be strengthened by the incorporation of other climatic and botanical factors, such as available plant moisture, patterns of radiation, day length and temperature.

In general terms, this hypothesis suggests that those farming systems that are situated in the semi-arid tropics and have as the major crops one planting of millet or sorghum will be most vulnerable to labour loss as a result of AIDS. Many farming systems of the semi-arid tropics include pastoral activities, which are not characterized by marked labour peaks, although they tend to occur in the dry season where herding to more distant pastures and watering places becomes necessary. While reported cases of AIDS have suggested a low rate of infection in the drier areas of the Sahelian zone (for example, Ethiopia, Nigeria, Chad, Somalia and the Sudan - all, incidentally, countries where accurate reporting is most likely to be difficult), the farming systems of southern and eastern Africa in Tanzania, Zimbabwe, Kenya, Botswana and Zambia also seem particularly vulnerable, since seropositivity rates have already reached high levels in some rural areas of those countries. In all cases there will probably be implications for foreign-exchange earnings and urban food supplies. It is in these systems that further studies should be initiated using existing secondary data.

In conclusion, it is as well to re-emphasize the difficulties of predicting the impact of AIDS at the system level. We have seen elsewhere in this book, as well as in studies that were able to use farm-management data at the household level, the range of adaptive mechanisms to labour loss and that these will be reflected at the farming-system level. This range includes such responses as intercropping, increasing the length of the working day, staggering agricultural activities (with important but varying costs in terms of reduced yields), new labour-sharing arrangements (maybe involving new household structures), switches to other activities such as hunting and gathering, charcoal burning, palm tapping, etc., and a reduction in the consumption of certain foods and rural-urban migration. All of these adaptations will tend to reduce incomes and the ability to self-provision. The precise patterns of these outcomes can only be predicted in outline. Instead of attempting more sophistication in the techniques of prediction, which undoubtedly bring professional rewards in terms of rigour and elegance, it may be useful to develop cheap, robust methods of monitoring the impact as it occurs. These can operate together with the broad predictive modelling being developed by the World Bank and the Food and Agricultural Organization along the lines of the methodologies developed in earlier publications (Abel et al., 1988) and in this book.

A note on further reading

The most easily accessible publications on the issues raised in this chapter are: Abel et al. 1988 and Gillespie 1989b.

9

The Way Ahead – AIDS in Africa: the Wider Picture, Current Responses and the Future

The wider picture

By April 1991, the World Health Organization reported that there had been a total of 90,646 cases of people with AIDS in Africa. The numbers who are currently HIV+ in Africa are unknown (although it is estimated that between nine and eleven million people are HIV+ worldwide); the difficulties of making such estimates are very great (Berkley et al., 1988, pp. 79–85); but the figure may be in excess of six million (Brown and Concar, 1991). At the International AIDS Conference in Florence, in June 1991, President Museveni of Uganda described Africa as facing an 'apocalypse' with millions dying and economies and societies collapsing (The *Guardian*, 17 June 1991, p. 1). Even without the AIDS pandemic, in 1991 the future looks grim for many of the peoples of Africa. Some have suggested that the continent can look forward to the new century preceded by at least a decade of permanent emergency – war, disease, food shortages, large-scale population movements, the maldistribution of food, famine and debt as well as the effects of swingeing structural adjustment programmes. The entire question of AIDS in Uganda and in Africa must thus be seen in the broader context of the problems that assail Africa as a whole.

Africa is a rich continent, with a history reflecting the ambiguous impact of colonialism, decades of post-independence bad government, civil and regional wars, continuing unequal trade relations with the world economy, the dabblings by external powers in national and regional politics and the debt burden that has resulted from all these things. In 1988, the World Bank projected a growth rate of GDP from zero to 0.7 per cent for the countries of sub-Saharan Africa between 1987 and 1995. However, low as these figures may have seemed, they are based on a preceding period when GDP actually declined across the continent. Between 1980 and 1987, these countries registered a decline of 2.9 per cent per year in GDP, compared with a rise of 1.8 per cent for all developing countries in the same period, and of 5.5 per cent for low-income developing countries (International Bank for Reconstruction and Development, 1988.)

Even the optimists from within Africa see the situation becoming worse before it gets better – and both individuals and governments find it difficult to take seriously the addition of AIDS to the plethora of other problems. By mid–1991, the food situation on the continent was such that Ethiopia, Sudan, Angola, Mozambique, Somalia and Liberia required $1 billion-worth of food to help 27 million people: the numbers are staggering and the food was not, by and large, being provided by the traditional donors. However, these are not the only countries facing a critical situation. Many other countries are also at risk. In the event of a failure of the rains for yet another year, Chad, Niger, Burkina Faso, Mali and Mauritania will be in the same extremity as these other countries (see Figure 9.1).

The 1980s saw average per-capita food supplies in sub-Saharan Africa fall by around 9 per cent, while reported (and the statistical reporting systems are very variable) mean gross national product fell by almost 3 per cent. In some parts of the continent, the population continues to increase. Between 1985 and 1995, it is estimated (without AIDS) that the continent's population will almost have doubled – from 210 million to 405 million, resulting in a food deficit equivalent to 15 million tonnes of maize. Even if the AIDS pandemic were significantly to reduce this rate of population growth in the longer term, as some current research suggests, its specific impact, removing the most productive rural age groups, is unlikely to make the balance of population to food supply more favourable. The loss of agricultural producers can hardly be expected to make up for the kind of food deficit that currently confronts the people of Africa.

Even less likely is an adequate response to this situation by countries in many of which government and the state appear to be in serious decline. While President Museveni's government attempts to reconstruct a state and a society seriously disrupted by Amin and Obote, and while Presidents Kaunda of Zambia, Bongo of Gabon, Mobutu of Zaïre and others face popular moves to alter the nature of government in their countries, Sudan and Somalia are examples of something quite new. These are areas of Africa in which the disintegration of the state is already well advanced. Long-term declines in rainfall (right across the Sahelian belt), the prevalence of war, the resultant breakdown of civil society, the movements of population – huge refugee populations moving within countries and between countries (Figure 9.1) – all contribute a volatile and distressing human background to the pandemic and a depressing picture of disorder in which population movement, the breakdown of established sexual customs and further declines in the economic status of women are bound to occur.

Wider world political changes, the collapse of Soviet power in particular, also mean that investment and aid is less likely to flow towards Africa. The first half of 1991 may provide a warning of what Africa can expect from the world community in response to the impact of AIDS. The aftermath of the Gulf War, Kurdish refugees, Shi'ite refugees, a cyclone in

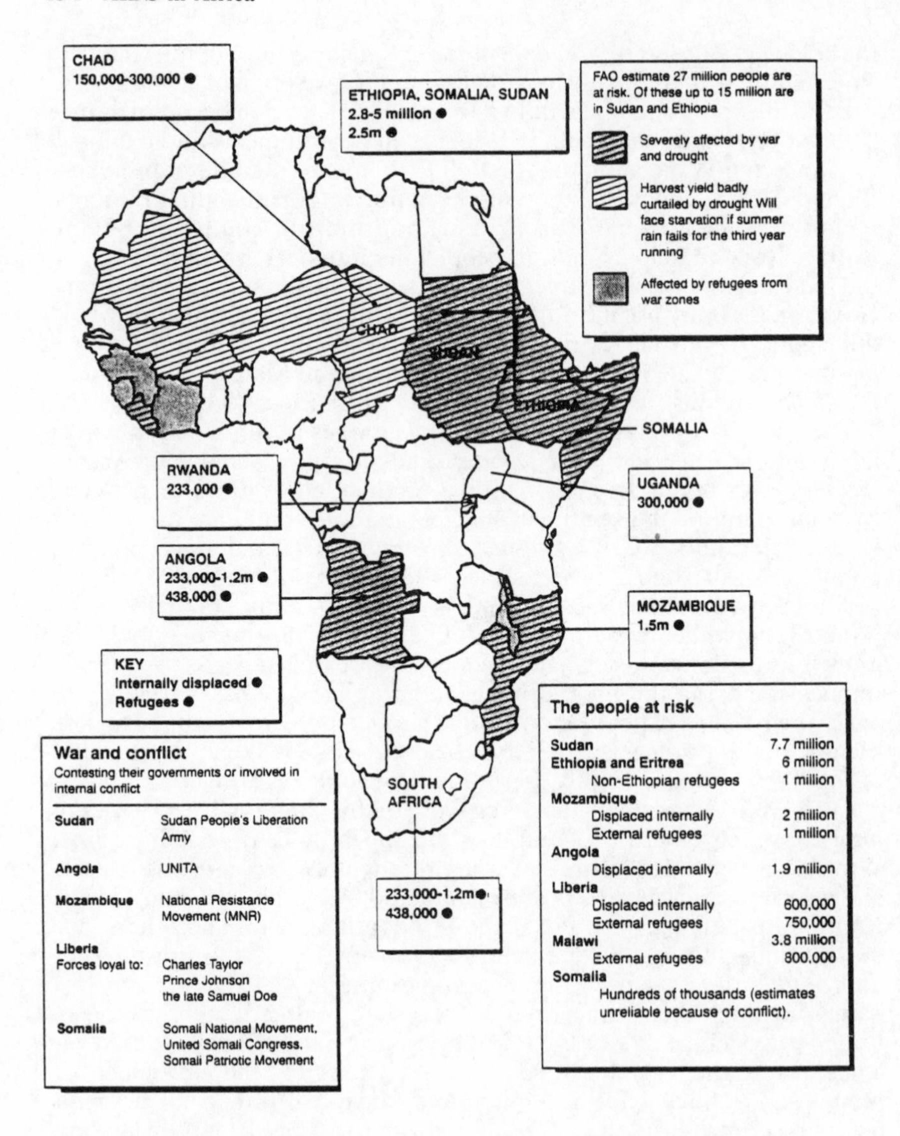

Figure 9.1 Drought, famine, refugees and warfare in Africa

Sources: Various, including The *Guardian*, 26 April 1991, p. 24.

Bangladesh, a cholera epidemic in Peru – the result may not only be 'donor fatigue' but also 'donor despair'.

Current Responses

Faced with such a depressing picture, all is not, however, pessimistic. The evidence from this account of the effects of the epidemic in Uganda is that realistic engagement with the long-term effects of the AIDS epidemic must take account of the creative responses from households, communities and governments – in combination, and that in the end, the main channels for coping with the epidemic may not be governments at all. Rather, communities and local and international NGOs may be the main channels through which support for affected individuals and communities is provided. They, and those who would lend support, can only work with what is available, within the framework of responses already being developed by Africans. And there are, as has been seen in Uganda, already many such responses, albeit faltering and starved of resources. They are not, of course, restricted to Uganda. Elsewhere in Africa, at Chikankata in Zambia (Williams, 1990; Campbell and Williams, 1990) the work of the Salvation Army Hospital is an example of one response to the epidemic, although, it must be emphasized, a response that is not necessarily replicable elsewhere given the unusually high-level administrative organization that characterizes this project and that might not be available elsewhere.

The aim of the project at Chikankata is to cope with the effects of the epidemic, recognizing that there is for the present no cure, but that people with AIDS or people who are HIV+ and their families need care, and that communities in heavily affected areas require some structure of care if the fear elicited by the disease is not to give way to despair. The strategy developed at Chikankata is based on the family and the household, the two units which, as we have seen from Uganda, provide the main care and undertake the lion's share of coping. The model is not necessarily replicable in other places, but its principles tell us something of how the problems can be approached. The two organizing principles of the Chikankata approach are integrated care for people who are HIV+ or who have AIDS, together with attempts to prevent further infection both within the community and among the health workers.

We have suggested that the main mechanism for confronting the epidemic may have to be based on cooperation between NGOs (both local and foreign) and the local communities. Chikankata shows this kind of cooperation in action. The origins of the system developed here lie in an offer from a foreign NGO to provide funding for the establishment of a hospice for the care of those suffering from AIDS. However, the hospital recognized that this was inappropriate for the circumstances in Zambia. Firstly because a hospice would soon be overwhelmed by the numbers of patients who would, given the scale of the epidemic, require care. Secondly, existing systems of caring for the sick in the community provided an alternative, and it was argued, more appropriate and lower-cost model of care for an African environment. For these reasons, the first goal was to provide, as far as possible, home-based care – using the network of kin as the main means of caring for people with AIDS. However, in so far as networks of kin would not necessarily have the resources to deal with a very sick person, it was soon apparent that the services of the hospital would have to be extended away from the hospital itself. Patients would have to be visited in their homes, and thus a mobile support system would have to be established. However, the project did not aim to provide only nursing and medical care, but also emotional support and counselling – and all in a rural area, with its characteristic problems of transport and communications.

Once this major principle of extended home-based care had been established, the second principle came into play – the prevention of the spread of infection through contact-tracing, education and community counselling. This approach has been successful. The hospital now provides in-patient care for people with AIDS, mainly on a short-term basis (the average length of stay is about 16 days), out-patient care through its mobile teams in the area, performs HIV testing, counsels people with AIDS and their families and also works with communities to educate them about ways to reduce the rate at which the disease is spreading. Thus the hospital deals with the issues of both care and prevention. The impact has been considerable; in some cases, community leaders have come to the hospital to request support for their own initiatives in AIDS prevention at the local level.

A positive lesson that might be learned from this experience is the importance of working with what the community already has: a lesson which, we believe, is also apparent from the present account of Uganda. This approach builds upon the existing strengths of the kinship networks. It tries to use customary views of sexual morality as part of AIDS education; and where custom might increase the risk of infection, it has tried to adapt custom. Nowhere is this more important than with the custom of the ritual cleansing of a widow or widower through sexual intercourse with a member of the dead person's family. Variants of this practice are found in a number of societies (including among the Banyankole people who live in parts of Rakai and formed part of the study population described in this book) and it clearly has a potential for spreading the disease. The response from the team at Chikankata has been to make efforts to persuade people to adapt their customs in such a way that symbolic action replaces intercourse. There is, however, a serious danger with home-based systems of care such as the one developed at Chikankata. This is the danger that in handing back the caring burden to the household and the local community, their coping capabilities may be overloaded in the face of this major epidemic. In some respects, these important and imaginative NGO and community responses must be seen as creative, but essentially short term. There is bound to come a time when they will require much more substantial funding than is currently available.

In Zambia, another line of attack has been through the school system. One hundred and fifty Anti-AIDS Clubs have been established: their aim is to get young people, especially boys, to adapt their personal behaviour in ways that take account of the AIDS epidemic. Members of the clubs promise to avoid sex before marriage, communicate information about the disease and care for people who are ill with it. Interestingly, in the light of the material presented in Chapter 3 on the attitudes of Ugandan schoolchildren, the boys involved in these clubs have adopted a moralistic attitude to the epidemic: an attitude that has meant they do not accept as one of its aims support for the use of condoms. Dr Kristine Baker, who was instrumental in starting these clubs, says of this:

> I raised the issue of condoms but the boys themselves turned down any reference to it. They felt that any information about condoms reducing HIV transmission might function as an 'escape' clause for members, and the pledge would lose its strength and meaning. Although some people worried that the pledge was too moralistic, that's the way the boys wanted it. (Baker, 1989, p. viii)

Also in Zambia, there has been a novel departure in AIDS education through an NGO called the Copperbelt Health Education Project (CHEP). CHEP serializes AIDS and STD information once a week in the two daily national newspapers. It also produces radio programmes in six national languages, as well as using other, simpler but very effective means of communication – metal hoardings and notices on litter bins, notices on the backs of school exercise books, leaflets distributed with water bills and stands at agricultural shows (Mouli, 1989, p. 4). This Zambian NGO has also organized song contests with AIDS as the theme and the use of traditional theatre. Its slogans are instructive, a combination of stick and carrot – 'AIDS kills – avoid AIDS' and 'From today do not think about dying of AIDS, instead think about living with HIV.'

In Uganda itself, the most profound response to AIDS has been by TASO, The AIDS Support Organization (Hampton, 1990). Founded by a group of volunteers in 1987, this NGO provides counselling, education, nursing and medical care as well as some limited material assistance to people with AIDS and their families. It tries to involve people who are HIV+ or who are ill in its activities as workers and not solely as beneficiaries, thus operating within its principle of 'living positively with AIDS'. And the importance of this perspective is extended to conscious attempts to define the experience of AIDS positively as part of the coping process. Thus the word AIDS is rarely used within the organization – instead people with AIDS or who are HIV+ are referred to as being 'body positive' and are not described as patients but as 'clients' – of whom there are now over 2,000. In contrast to the Chikankata model, TASO is currently mainly concerned with urban and peri-urban people. At present it does not have the resources to provide care for rural populations, although with the establishment of offices in Mbarara and Tororo, as well as its original sites in Kampala and Masaka, such provision may soon become possible. However, although TASO has not been able to provide the kind of mobile, rural-oriented care that has been central to the Chikankata model, the Kitovu mission hospital, with clinics in Masaka and Rakai Districts of Uganda, has initiated such a programme. Both Kitovu and Nsambya hospitals (in the Masaka District and Kampala respectively) have mobile clinics. These mobile teams, visiting about 40 people per week, work through existing community networks. Local leaders identify people in need of treatment and help and act as a referral system. Where communities are not open to such assistance, efforts are made by these mobile teams to provide some basic AIDS education. (WorldAIDS, July 1989, p. 8).

TASO is an important example of a Ugandan initiative which, from its inception, has been closely tied to the development of ideas about coping with AIDS and HIV in the UK and the USA. Its founder, Noerine Kaleeba, had to cope with the personal impact of the disease when her husband became ill in Britain in 1986. She became part of a support network in the UK, took this experience back to Uganda with her and founded TASO. Built into the development of this organization has been a recognition that it would take the best of what is available from the experience of UK (some of its counsellors have been trained there) but that this must be combined with the strengths available from within Ugandan culture and society. This has required a deal of experimentation and innovation as well as hard work to raise funds and organize what is now probably the most important grassroots response to AIDS in Uganda outside the official medical and research services. The important principles it teaches are those of living positively with the disease and its implications and also that people with AIDS or who are HIV+ do not thus become removed from society: they remain an important part of it, able to perform their roles as carers and producers. They can and do help themselves and each other; they can and do continue as valuable and valued contributors to their societies. This was recognized by Noerine Kaleeba at a Swedish award ceremony in recognition of her work, when she said:

> I wish there was more than one award because a lot of people in TASO deserve to receive it. One person who encourages me daily is a woman who is HIV, like her husband and her three children. She makes a tremendous contribution as a counsellor. If I were in her shoes, I wonder if I could do it. (WorldAIDS, January 1991, p. 6).

This, perhaps, is the essence of community coping with the pandemic.

In addition to these types of response that focus on those who are ill or who are HIV+, there have also been other initiatives aimed at prevention and coping with the downstream impacts. In Chapters 6 and 7, we described something of the work of two Ugandan NGOs which are attempting to do something about the orphan problem. But efforts are also being made in the field of education and prevention. Action for Development, a women's NGO, visits schools and talks to girls about the dangers of contracting the disease. Given the economically disadvantaged position of women in general in Uganda, it is the poorest young women who are the most vulnerable (Watson, 1989, p. x). What we see in the work of TASO, the Orphan Community Based Organization, the Ugandan Women's Efforts to Save Orphans, Action for Development, as well as several other Ugandan NGOS, is the often halting, certainly resource-starved, bravely experimental efforts by Ugandans to cope with the impact of the pandemic in their own way.

Similar attempts are under way in other parts of Africa. In Nairobi, the Amani Counselling Service, founded in 1979 to help people with a range of personal problems, has been offering AIDS counselling since 1988. However, there are also important interchanges of experience within Africa. Thus the Kenya AIDS Support Organization, established in mid-1990 along the same lines as Uganda's TASO, offers counselling services and is run by people with AIDS and who are HIV+, as well as others.

Regional cooperation is apparent in Southern Africa (WorldAIDS, November 1990, pp. 5–9), where, after a period of denial and refusal to recognize the potential seriousness of the epidemic in some countries, attempts are now being made by both governments and NGOs to confront the issue. The founding of a regional NGO coordinating and liaison organization, the Southern African Network of AIDS Service Organizations (SANASO) at a conference in Harare, Zimbabwe, in 1989, was an important milestone. There is a particular need for effective action in this region, given the existence of widespread long-range labour migration as well as the extent of disorder associated with the struggle against apartheid. Warfare and population movement, as we saw in Chapter 5, provide fertile conditions for the spread of the disease.

Across this region, NGOs are taking initiatives. In Lesotho, the Lesotho Red Cross and other NGOs are working closely with the National AIDS Control Programme. In particular, there is a scheme run by the Lesotho Red Cross whereby 3,000 registered Red Cross Youth are being trained to educate their peers about the nature and danger of AIDS. In Swaziland, the National AIDS Control Programme includes NGOs as well as traditional healers. Under its auspices, there have been several regional meetings with women and girls to discuss controversial issues such as condom use. Swaziland is the only country in Southern Africa which has currently produced educational material especially for men. In Zimbabwe, the National AIDS Control Programme is also working closely with NGOs and with traditional healers. In particular, the Harare AIDS Counselling Trust has sponsored other NGOs, including the Women and AIDS Support Network aimed at protecting younger women from infection. The NACP has enlisted the help of traditional healers and has trained some of them as AIDS counsellors. Even in Mozambique, racked by years of civil war with its associated economic and social disruption, efforts are being made to confront the disease. With assistance from a Swedish NGO, young people, the military and STD patients, as well as midwives and traditional healers, have all been enlisted in efforts to educate and inform the population.

South Africa confronts particular problems as the disease spreads in a society experiencing rapid political change against a background of apartheid. A major problem has been to enlist the black population in the fight against AIDS at a time when black people are, understandably, suspicious about any initiatives from the white minority government. The issue of condoms, for example, which is always problematic, here takes on an added dimension as some blacks interpret advice as to their use as a ploy to reduce the rate of black population growth. Despite this, the foundation of an AIDS Task Force under the auspices of the African National Congress in 1990 (Whiteside and van Niftrik, 1990, pp. 1–2), intended to challenge the state's inadequate response, draws on international support from foreign NGOs, the WHO and also enlists the support of trusted local NGOs and the churches in particular. The Maputo Declaration under which the AIDS Task Force was established is an attempt at a comprehensive approach to tackle the problem of AIDS in South Africa and also a clear public indication of the ANC's commitment to fighting the disease. However, the problem of AIDS in the republic is a very particular one as the minority white government and the ANC, inexperienced in internal adminstration and government, struggle towards some kind of effective response. Even before the Maputo declaration, the range of responses at the community level had been very broad. Thus, a non-racial hospice and commune for people with AIDS was established in Capetown in 1989. A small beginning, initially accommodating only 12 people, this facility is open to any HIV+ person and has links with a hospice in Khayelitsha about 40 kilometres away where people with AIDS are cared for by nuns of Mother Teresa's order. In Soweto, the Township AIDS Project (TAP) was first conceived in 1986 by members of the local community. Since then, TAP has worked with existing groups in Soweto educating and sensitizing them about the problem. In particular, TAP has been making efforts to reach men who have fewer accessible organizations than women – through the illegal bars, shebeens and through groups associated with football. However, here as elsewhere, the issue of condom use has met with considerable resistance.

Condoms and condom use are among the most difficult areas of any programme designed to slow down the spread of AIDS. This question impinges upon people's expression of their sexuality, on gender relations, on their religious beliefs and also, as we have seen in the case of South Africa, may easily take on a political complexion. Cultural attitudes towards condom use, particularly among men, may make them very difficult to accept. They are seen as wasting time, wasting sperm, preventing flesh to flesh contact or even belittling the sperm by obstructing its entry into a partner (Serote, 1990, pp. 5–6). Religious objections to condoms, particularly in countries with large Roman Catholic populations, are also of importance, for example in Uganda where 50 per cent of practising Christians are

Roman Catholic. Thus President Museveni has adopted an ambivalent position on the question. In some accounts he was apparently so impressed by modelling predictions that Uganda's population estimates for the year 2010 could be reduced from 37 million to 20 million by AIDS that whereas the government's AIDS Control Programme had previously stressed sexual abstinence and faithfulness as the main ways of combatting the disease, President Museveni reluctantly but publicly announced that Ugandans should use condoms to protect themselves (Tebere, 1991, p. 3). However, as recently as June 1991, at the Seventh International Conference on AIDS, he stated: 'I don't support this idea of condoms myself', suggesting instead that 'abstinence and self-control must be taught to our young people' (Brown and Concar, 1991). In Zambia there has also been conflict between the churches and the government over the issue of condoms and the messages of health educators. In 1988 the Ministry of Health released a booklet for secondary school children suggesting ways (including condom use) in which they could avoid infection. This publication was attacked as 'irresponsible and immoral' by a Protestant church leader. To his credit, and reinforcing the relatively open attitude of the Zambian government, President Kaunda responded positively by stating that it was the duty of the party and the government to talk publicly about AIDS and how it could be combatted (Lutunda and Kabwe, 1989, p. 5).

But in addition to problems of attitudes and concerns about sexual enjoyment, perhaps the most important difficulty with condom use has to do with poverty and the fact that their use is dependent upon male cooperation. Crowded living conditions, the lack of running water and the cost of the condoms themselves all present obstacles to their easy acceptance. In Kenya, condom use appears to be on the increase. Village chiefs, in association with the Kenyan Family Planning Association, have been campaigning for condom use to combat HIV infection. But even so, the following case-study should focus our attention on the realities of condom use in rural Africa.

Miriam and her husband find it difficult to comply with the community distributor's advice to inspect each condom for possible tears before use. They light lamps in the night only if the baby is sick and in need of attention. Miriam accepts using condoms, but she feels that some of its demands do not take account of the realities of rural life – like keeping enough water available for washing the hands after removing the condom. She says: 'My children are still too young to help me fetch water. The well is not so near, and fetching water is not the only job there is to do. There is firewood to be collected and food to be prepared. Perhaps I don't need so much water for my hands but I don't like this new use for water.' (Munyako, 1991, p. 3).

In addition, the problem of condom disposal in a crowded environment is not to be underestimated – particularly in rural environments where there may not be pit latrines. All this provides an additional burden for men and women in circumstances where life is already difficult.

Even so, there is some evidence that condom use may be on the increase. At the Fifth International Conference on AIDS in Africa held in Kinshasa, in October 1990, many speakers from African countries reported the increased acceptance of condoms by both men and women at high risk. A striking example came from Zaïre where a programme sponsored by the United States agency Population Services International had apparently resulted in the number of condoms being distributed rising from 200,000 in 1986 to a projected nine or ten million in 1990 (WorldAIDS, January 1991, p. 3; Brown, 1990, p. 41).

Women in particular confront the difficult task of persuading their husbands or partners to use condoms in situations where women's economic dependence upon men is high and acceptance of men having multiple sexual partners is the reluctant norm for many women. Their dilemma is apparent from the following report from Zimbabwe:

> Norah, one of my friends, asked me for advice one day. She was worried that her husband might be infected with HIV and pass it on to her. She knew that her husband often 'met with' girlfriends at beer halls and hotels. She has never liked the fact but regarded it as inevitable. She had never mentioned to him that she knew about his girlfriends because this would make him angry.
>
> I asked her whether she had thought of suggesting that they use condoms. Of course, she told me, but even the suggestion would cause problems. Her husband would either accuse her of infidelity, or charge her with accusing him of infidelity. He would become angry and not use the condom. If she asked him to use condoms when he went with girlfriends she would reveal that she knew about them. This again would make him very angry. Had she talked with her husband, in general terms, about the problems of AIDS in the community, I asked. She said that she had tried to tell her husband how serious the disease was, but, like other men, he did not believe that AIDS existed in the area. We finally agreed that all she could do was to try to talk to her husband once again about how common HIV infection is in the community, using a story about an imaginary friend who has been infected with HIV by her husband.
>
> Norah loves her husband and does not want to leave him. Her children need their father and the family needs money for school fees, books and clothes. Her fears that he might contract HIV and infect her are well founded. She is a well-educated, mature woman in a responsible skilled job, but as the balance of power in her marriage lies with her husband, she is prevented from protecting herself from HIV. (Holmes, 1991, p. 9)

So, despite efforts to persuade people to use condoms, despite the import of millions of condoms to Africa (nine million to Zaïre in 1990, two and a half million to Mozambique in 1991 (*AIDS Analysis Africa*, Vol.1, No. 2, May/June 1991, p. 6), it is millions of individual decisions which will decide the effectiveness of a programme of condom use. The failure of vending machines in Botswana (*AIDS Analysis Africa*, May/June 1991, p. 6) rumours that the USA is providing sub-standard condoms to Zimbabwe, beliefs that condoms are part of the South African government's efforts to reduce the rate of black population growth, or that condom use is all part of some other kind of conspiracy against Africans, or simply men's unwillingness to recognize the risk and the issue of trust between partners in a steady relationship (Brown, May/June 1991, p. 5) – all these factors make condom programmes difficult to introduce.

However, perhaps more important than all of these factors is the difficulty of using condoms in circumstances of poverty and overcrowding. Condoms require a degree of forethought and sexual leisure time which is, possibly, not available to the poor of Africa's cities and countryside. The difficulty of introducing them as part of family-planning programmes in many parts of the world should be a warning to anybody who thinks that they are a simple, straightforward solution – not least of all because of the political issues that may surround the entire question of their introduction and use.

However, it is, perhaps, the position of women and the relationship between women and men, which is a central issue in the fight against the spread of AIDS – not only in Africa but also more widely. Throughout this book, we have argued that the social and economic position of women, the social and economic context of the relationship between men and women, is a crucial variable in understanding the way in which the AIDS epidemic has developed in Uganda. Economic insecurity, land-tenure law and custom have all contributed to a particular pattern of heterosexual spread. We say a *particular* pattern of heterosexual spread because there are clearly other patterns of heterosexual spread that do not arise from the specific combination of factors within which women's vulnerability is constructed in Uganda. However, it is certainly the case that, throughout Africa, the ways in which women relate to the disease, their knowledge of it, their choices of sexual partner and their ability to say 'no', are all of the greatest significance. We have emphasized the economic, political and social position of women as a major issue in this book. We have done this because we believe that attention to these issues is of the greatest importance in practical engagement with the nature and implications of the epidemic. It could be argued that such a view fails to appreciate the position of women in a specifically 'African civilization' (Caldwell, Caldwell and Quiggin, 1989, pp. 185–234). However, while we must be careful not to make moral evaluations across cultures – and in this book the term 'promiscuity' has been studiously avoided – it is not constructive to argue that judgements must never be made about the implications of behaviours in cultures other than our own. Caldwell and his collaborators argue that campaigns aimed at affecting the position of women in Africa may 'severely undermine social institutions, hurting in particular the position of women and some of the most socially marginal groups' (Caldwell et al., 1989, pp. 224–5). However, while such a view merits consideration in the caveat it offers about hasty judgement or action, it also fails to take account of certain very important aspects of the situation of women in contemporary Africa.

The first of these is that, in circumstances such as those we have described in Buganda, women's position is no longer 'traditional'. Indeed, the characterization 'African civilization' assumes a social and cultural stasis in Africa which does not reflect the immense changes that have been occurring there and that continue to occur. Secondly, defences against criticism of an assumed static 'culture' fail to recognize that 'culture' is also ideology – a set of ideas that defends the social and economic position of one group at the expense of another. In so far as a culture privileges men over women, as is the case in most societies, not only those of Africa, there may be characteristics that merit criticism when the result of that privileging is to place a whole category of the population – women – in a vulnerable position *vis-à-vis* men. And thirdly, and most importantly, discussion of the nature of 'African civilisation' without reference to the realities of contemporary social and economic relations does little to enable people to cope with a very serious set of circumstances such as those now existing on the African continent.

The campaign by the Ugandan NGO Action for Development (ACFODE) (Watson, 1989, p. x) confronts the issue very frankly indeed, recognizing the ways in which poorer young women in a very poor country are vulnerable to the blandishments of older men who can offer them economic rewards as well as emotional and sexual support. One of the workers with this project puts the point nicely when she says of the advice that the project's counsellors offer:

> We tell them they should look after themselves as individual and important people. They should not look at boys as their future. They should stick to their education and forget about sex. (Watson, 1989, p. x)

This problem has been widely and publicly discussed in Uganda and President Museveni has called for legislation to protect children and minors (defined as those under 14 years of age) against infection. At a seminar in 1989, he called for legislation to make it a severely punishable offence for an adult to seduce a minor. Such legal provision may be important in a society in which some men are seeking out younger women in the belief that they are less likely to be infected. (Dyakire, 1989, p. 4)

However, the problem is not restricted to younger unmarried women. Women who are married or in long-term relationships are also very vulnerable; for them saying 'no' is neither easy and nor does it fit with the traditional expectations of men and women in their relationships. Such problems are not restricted to Africa, as the recent changes in British law regarding the possibility of rape within marriage show. However, the problem is certainly no less important in Africa. One vignette will serve to further illustrate the point. In conversations with a group of women school teachers in a village near Machakos in Kenya, it became apparent that despite their comparatively high levels of education and their knowledge of the dangers of AIDS, these women could neither say 'no' nor discuss condoms with their husbands – a difficulty that was of the greatest importance given that each of the women in the group was married to a man who spent weeks working away in Mombasa. These women felt that asking their husbands to use condoms or discuss the AIDS problem would threaten their marriages.

Such economic and social vulnerability confronts the broader question of the legal status of women. This has been seen in the present study in relation to women's rights to land in Buganda. The question of women's legal status and its relationship to the epidemic has been discussed openly in Tanzania. A seminar held in 1990 concluded that discriminatory laws on inheritance and succession mean that most women whose husbands have died from AIDS suffer financial hardship. The 1963 Customary Law of Succession prohibits women from inheriting property while sons and brothers of their deceased husbands are alive. This legal disability is reinforced by social attitudes that blame a woman for her husband's death, accusing her of not having cared for him properly. In such circumstances, his relatives may cheat her of his property and expel her from the marital home. In addition, if mothers and their children are also infected by the virus and fall ill soon after the husband's death, 'it is almost impossible for widows to navigate the complex legal procedures to reclaim their property'. Although the problem is now recognized in Tanzania, government responses have been slow. In 1989 it was suggested that a parliamentary commission be established to investigate the legal and political aspects of AIDS. It has yet to be convened (Ogola, 1991, p. 5).

Women remain vulnerable in Tanzania, as elsewhere, not only because of the legal and customary framework of their lives, but also because of the attitudes of men whose identities are in turn closely connected to their perceptions of their sexuality. Thus, in Dar es Salaam, the younger men are reported to have interpreted the initial letters of the disease in a way that actually reinforces their existing sexual attitudes: the acronym AIDS has been converted into the KiSwahili phrase 'Acha Iniue Dogedego Siachi' – 'let it kill me as I will never abandon the young ladies'. Similar attempts to pass off the threat of the disease by reinterpreting the acronym have been seen in many other countries. For example in Cote d'Ivoire, the French acronym, SIDA, has been translated as 'Syndrome Imaginaire pour Décourager les Amoureux' (Imaginary Syndrome for discouraging lovers; Kouassi, 1989, p. 4). And more widely, women say that: 'We are afraid of AIDS, but very few men want to use condoms; if we develop AIDS, it will be in the line of duty,' and 'Even if I take precautions, how can I be completely safe while I have no control over so many things – including my own husband?' (Mmbaga and Ogola, 1989, p. 3).

However, even the recognition of women's vulnerability to infection can have a dual impact, placing the majority of the responsibility on women rather than bringing the responsibility clearly home to men. Thus even the well-intentioned actions of government can carry unanticipated messages. In 1990 the Acting Principal Secretary of the Ministry of Health in Lesotho, Mr Tehoho Kitleli, made the point publicly that the economy of Lesotho was in the hands of Basotho women and girls. He said that if this group were to die of AIDS, the whole economy would be severely affected. While recognizing that women were often victims of rape or drunken sexual advances, he also stressed that women should learn and reinforce the safe sexual practices that might contribute to slowing the spread of AIDS (*Lesotho Today*, Maseru, quoted in *AIDS Analysis Africa*, Southern Africa Edition, February-March 1991, p. 11). While it is true that women do need to know about AIDS, this knowledge is of little use to them in societies where they continue to be disadvantaged through custom, through law or through both and are thus unable to take any effective action to change their lives.

Levels of response to the epidemic by government vary between countries. There has often been a reluctance to accept the inconvenient implication of the disease. This kind of response is not, it must be emphasized, restricted to the countries of Africa. In the UK, in 1989 the then prime minister, Margaret Thatcher, stopped government funding for a survey of sexual attitudes in the general population, considering that the collection of such information was intrusive and unnecessary (*AIDS Newsletter*, October 1989, p. 2). And in any case, the costs of dealing with these implications are clearly beyond the economic resources of most of the countries of Africa. Sometimes, indeed, the very nature of governments may themselves be a factor in the form of the epidemic. To take one example: many African countries have military governments. One count suggests that of the 15 countries of West Africa, 11 are led by members of the military (Lyerly, Barry and Miller, 1990, pp. 1 and 8–10) and in many other African countries, the military is vital to the preservation of internal order, national integrity or merely the survival of a particular regime. Military populations are mobile, consisting of young men; in some central African states they are reported to be over 50 per cent seropositive. In Zimbabwe it was rumoured in 1989 that 'up to half of the 46,000 strong national army is HIV positive' (*Africa Analysis*, 2 February 1990, p. 2). If these levels of seropositivity are any reflection of the general situation among the armed forces of Africa, then two things follow. The first is that armies themselves may well be important disease vectors; the second is that the very basis of state power in some African countries is threatened as military personnel fall ill and die.

Most countries in Africa have established the basis of a programme, in association with the WHO, to begin coping with the epidemic. However, political constraints, lack of resources and, in some cases, the sheer inability of the state to respond, have meant that the responses vary greatly from country to country. Opinions on the effectiveness or otherwise of government responses will vary depending upon the position of the observer.

A central argument of this book has been that we should not be treating AIDS as though it were just any other disease – although the illness and death resulting from malaria, tuberculosis, diarrhoea, bilharzia and all the other diseases that cause hundreds of thousands of deaths or stunt millions of lives on the continent should not be discounted. What is different about AIDS is that it threatens to reduce the size of the productive cohorts of Africa's population; it will cost raw labour power so essential to agricultural production; it will cost skilled labour power trained at the expense of scarce resources; it will strain health budgets and it will decimate administrations. In this book, we have indicated the ways in which the epidemic might affect agricultural production. A comment by somebody writing from an agricultural research station in central Africa brings the point home in a particularly poignant way. He says:

> In our programme we have had major success by introducing climbing beans into non-traditional areas of the country in replacement of bush beans. Climbers have higher productivity per unit of land but require more labour input. Of course the AIDS situation is of interest to us in this respect. AIDS may make climbing beans an irrelevant technology within the next couple of years. You may know the AIDS figures for this country . . . and I can tell you that it is true. We see people dying, and many, too many. It is a disaster. Up to now the authorities are claiming that it isn't all that bad in rural areas, but god knows that this is a matter of time. (Private communication, 1990)

Such results are not restricted to Africa. They are just more pronounced there because of the extreme poverty of the continent. Research in other parts of the world indicates that similar concern is being expressed in other poor countries, too, for example in Mexico, a country that is poor but immeasurably richer than most in Africa (Christenson, 1988, pp. 619–28). This message should be clear from what we have said about the situation in Uganda. Across Africa, evidence for the seriousness of these downstream effects is accumulating rapidly; given the nature of the disease and the shape of the epidemic curve (discussed in Chapter 2), now is the time to take action to mitigate the worst effects in the next two decades. Because this is a long wave disaster, we must repeat the point that the effects we are seeing now in Uganda and elsewhere are the result of events (personal, communal, regional, national and international) that occurred a decade or more ago. Action taken now cannot change the present, nor can it change the immediate future. It can change the way the situation will look in the years after 2010. The first socioeconomic effects of the myriad individual and communal decisions and events of a decade past can now be seen across the continent.

The medical costs alone are frightening. In Africa generally, health-care budgets are very low – cash available per head of the population varies from $1 in the very poorest countries to $10 per year in the better-off countries. These sums are inadequate. They are even more inadequate in the face of calls for structural adjustment. They allow little real medical care for populations whose health and nutrition status is already low, and certainly nothing for training staff and providing health care for public campaigns and personal treatment of people with AIDS. In sub-Saharan Africa, governments on average spend $3.50 per head on health care each year compared with Scandinavia where the amount is $1,000 (Brown and Concar, 1991). One HIV test costs around $1 and machines for mass testing of blood samples each costs as much as $15,000 (Feurstein and Lovel, 1989).

As an illustration of the costs of the epidemic to the health budgets of even the richer countries of Africa, the case of Nigeria is informative. There, as early as 1988, AIDS was the only disease to be earmarked for separate financial allocation (more than US$1.5m in 1988) outside the national health-care system. This sum was only exceeded by the immunization campaign against the six leading child killers (*WorldAIDS*, January 1989, p. 10). Preliminary attempts by health economists to estimate the direct costs of health care (the cost only of treating and caring for those who directly suffer from the disease) for each person with AIDS in Tanzania and Zaîre range from $100 to $1,500 (Over et al., 1988, pp. 123–35; Skitovsky and Over, 1988). It has been estimated that given the numbers of people who may be seeking treatment in the next decade, the total direct costs of AIDS in Zaïre alone may be as high as $49m by 2010, and that to finance these costs, the country's national budget would have to increase by 58 per cent in 1993 and further increase by another 244 per cent in 2010 (*Southern African Economist*, April/May 1989, pp. 53–4). And Zaïre is by no means unique in the effect that AIDS may have on its health infrastructure, as the case material from Uganda has illustrated.

But the situation is in fact worse than this might suggest, for such direct costs are only a fraction of the real costs to the economy, which will include the costs of lost production, lost skills and the cost of caring for orphans. These indirect and long-term costs – the realities of the downstream socioeconomic impact of AIDS in Africa – will take many different forms. These will be distributed unequally between different countries in Africa, depending on the structure of the economy and society. For example, in countries such as South Africa and Zambia, where the mining economies attract labour across large distances, both within the country as well as from within the broader region, the epidemic takes on a specific form. In South Africa, there is evidence that the disease is spreading among the single-status miners and within the communities adjoining the labour camps that are characteristic of that country. The South African Chamber of Mines reported high levels of HIV among black miners in 1989. Many of them were from adjoining countries, the majority reportedly from Malawi. Accurate information on HIV rates is difficult to obtain, but some sources within the mining industry were reporting that the rates might be doubling as rapidly as every eight months. It has been argued that the apartheid/migrant labour system provided excellent conditions for the spread of the disease in the southern Africa region. Population mobility, civil and regional unrest and warfare all furnish large-scale social and economic disruption with the movements of workers, refugees and military forces – a situation which is in some respects similar to the one we have described, albeit on a much smaller scale, as forming the backdrop to the particular form of the epidemic in Uganda.

Africa as a whole is characterized by many of the problems that contributed to the particular form the epidemic has taken in Buganda. In addition, however, in South Africa itself, there may be an additional factor. The 'homelands', with their very poor health services, rural unemployment and malnutrition, may themselves constitute a serious obstacle to the implementation of any effective programme of AIDS prevention or containment (Rees, 1989, p. 6). Recent political changes also mean that as many as 40,000 exiles may be returning in the next few years, often from neighbouring countries that already have high levels of seroprevalence. In general in South Africa, it is rapidly being recognized that the epidemic will have a greater and more detrimental impact on the economy and society than will be the case elsewhere. It is certainly the case that while most shades of political and social opinion are taking the issue very seriously, from the ANC through its Maputo declaration to the larger commercial and industrial conglomerates, there remain veins of influence that play down the importance of the epidemic and government responses remain mixed (*AIDS Analysis Africa*, May/June 1991, p. 1), although the South African Government has recently established an AIDS unit within its Department of Health and Population. In Swaziland, where the most recently available figures are for 1988 and which had reported only 14 cases by June of that year, and where the HIV rate is said to be around 3.5 per thousand (based on samples from 'low-risk groups' such as blood donors and other volunteers), it is reported that the doubling time of the epidemic is around eight months. There is concern there for the effects of the disease on the sugar and forestry industries where some reports from its workers indicate seropositivity levels at around ten per cent (*AIDS Analysis Africa*, May/June 1991, p. 7).

In Tanzania, the area most affected by the disease is Kagera in the north west of the country, close to the border with Rakai District in Uganda, the subject of the main body of the present study. This area is a major coffee producing zone (coffee is Tanzania's leading export). Before the epidemic, the region was relatively prosperous. Now one report says dramatically that today: 'it is a place of the dead. AIDS is rampant in Kagera, attacking and killing young men and women in their prime. Complete villages have been wiped out . . . leaving only the very young and the very old . . . The region is now said to have more than 20,000 AIDS orphans; and it is feared that in four years' time the figure will be over 50,000' (Rajab, May/June 1991, p. 9). In addition to the economic impact of lost coffee production, there is also fear here, as in other African countries such as Kenya, that publicity could affect the tourist industry, an important source of foreign exchange for hard-pressed African economies and an issue not to be underestimated in countries that are deeply in debt and urgently in need of foreign exchange. In Kenya where tourism is the largest source of foreign exchange, earning the country in excess of eight billion Kenyan shillings a year in 1989 (about £222m or $366m), in 1988 reports in the Western press about AIDS caused a decline in earnings from tourism by around 30 per cent.

In Malawi, it is reported that there may be as many as 500 new cases of AIDS each month (the most recent report for January 1990 indicated that around 7,160 cases had been notified). The best guess is that perhaps 20 per cent of sexually active adults in urban areas and 17 per cent in rural areas might be seropositive, suggesting that around 750,000 out of the country's total population of 8.5 million might be infected (*AIDS Analysis Africa*, March-April, 1991, p. 5). The implications of these levels of illness and of seropositivity for a country predominantly dependent upon rural production (and which also carries a burden of over 800,000 refugees from adjoining states) are suggestive of an impact similar to the one we have observed in Uganda. In neighbouring Zimbabwe, whose economy is somewhat less dependent on farming, the main employers's organization, the Confederation of Zimbabwean Industries (CZI), has called on the government to come to terms with the threat it perceives the disease to pose to the supply of young, skilled labour to the economy. Information is in short supply, but the point is that employers believe that published figures are inaccurate and out of date. The CZI claimed in 1990 that between 10 per cent and 20 per cent of the general population might be seropositive and that the incidence of HIV was doubling every ten months. This suggests that in ten years time as much as 90 per cent of the workforce could be dying of AIDS-related diseases (*Financial Gazette*, 6 July 1990).

An important issue to be tackled is that of information and the education of populations about the disease and how to combat it. It is curious that in countries where cigarettes and Coca Cola find their way into very remote areas, where items of fashion clothing may be worn by young men emerging from the meanest village house, knowledge of AIDS and the availability of condoms may be in short supply.

The techniques of persuasion and distribution employed by international companies are not, by and large, applied to problems of public health education. To put the point bluntly, the budgets and the techniques for marketing AIDS information are small and amateur by comparison with the efforts made to sell consumer goods. This is particularly odd in so far as the greatest market sector for all these goods is also the one whose lives are most threatened by the epidemic. In the centre of any national capital in Africa, large hoardings and even neon signs vaunt the merits of this or that beer, soda drink or airline; elsewhere in the same capital, at the headquarters of the National AIDS Control Programme, amateurish posters are exhibited as evidence of the best that international and national 'experts' can produce to combat the spread of AIDS. These posters also hang forlornly from the wall of an occasional bar or eating-house. They have become part of the furniture and little notice is taken of them. It is odd that nowhere is the full panoply of techniques and resources of the advertising industry applied to the fight against AIDS. As with so much that has been written about AIDS, the situation in Africa is no different from that elsewhere. The government of the UK, a rich country, has allocated just under £10 million to education of the public about the disease (*Health Education Authority*, 1991a, p. 7). In comparison, the amounts spent on advertising tobacco products by the tobacco industry is around £113 million (*Health Education Authority*, 1991b).

The efforts that are being made to educate the peoples of Africa about the risks they face and the nature of the illness are worthy. They are also under-resourced and thus less effective than they might be if they were informed by the knowledge, research and market-evaluation procedures used by the best advertising agents and supported by the finance available to Coca Cola. Instead, NGOs and small arms of government struggle with the task: a task that is often made more complex, as in the case of condoms, by the political and religious divisions that characterize so many of the countries of Africa.

Attempts are, however, made to get the messages over to the population. The messages are various – AIDS cannot be caught through casual social contact, condoms may protect you against the virus, large numbers of sexual partners increase your chances of contracting the disease, nobody is to 'blame', people with AIDS should not be ostracized, and so on. These messages and others are fed into the communal consciousness via the full range of techniques and channels, books, posters, adult education classes, the churches and the schools. In Kenya in 1989, the children's magazine *Pied Crow*, funded by CARE, WHO, UNICEF and SIDA, was used as a vehicle to get the message over to children of primary-school age. 800,000 copies of the magazine were distributed throughout the country. There was an instant response: as many as 80 letters a day were sent to the editorial offices, some of them from countries other than Kenya. The magazine found its way to some adult readers, eliciting a demand for information (Tuju, January 1990, p. 4).

In Uganda in 1989, the concert by the nationally famous singer Philly Bongoley Lutaya (Musoke and Tamalie, 1990, p. 8) brought the issue home to large numbers of urban people. Bongoley had said publicly in April 1988 that he was suffering from AIDS. His record *Alone* which told of his experiences with the illness, a film documentary *The Life and Times of Philly Bongoley*, together with talks and lectures, made clear what many were already admitting: that everyone, from celebrities to the most humble, was vulnerable to the disease. Soon after his farewell concert, Bongoley returned to his country of residence, Sweden, where he died within a few weeks.

In Nigeria, on the other side of the continent, popular song has been used as a vehicle for the message about condom use. This resulted from participation by two popular musicians, Sunny Ade and Onyeka Owenu at a UNICEF seminar aimed at encouraging artists to use their art to get across socially significant messages. The messages of the songs are indirect, ranging from discouraging young women from starting sexual relationships with the message 'If you love me, you go wait for me', through 'Let us love with care', to the franker 'I don begin dey use rubber now' (Ogunseitan, November 1989, p. 5). From another direction, 21 government ministers publicly submitted to HIV tests on prime-time television in order to encourage the wider population to have voluntary HIV tests – apparently with little impact.

In another west African state, Cote d'Ivoire, ministerial broadcasts outlining the dangers, twice-weekly radio programmes, magazines advertising condoms as well as posters and banners in schools and public places, have all been part of the effort to combat the epidemic (Kouassi, May 1989, p. 4).

From Kenya, analysis of the types of letter received in response to a series of radio programmes 'AIDS the facts', beginning in 1988, illustrates the issues that public education programmes confront. By November 1989, a total of 9,897 letters had been received and a random sample of one third of the letters revealed the types of concern shown by literate people in Kenya. These fell into nine clear categories:

Category 1: those who thought the disease was 'just propaganda', that it only affected 'bad' people, or that contracting it is 'just an accident' – like any other disease;

Category 2: those enquiring whether the disease can be caught by casual social contact such as sharing a bed, plate or cutlery with a person with AIDS;

Category 3: those enquiring whether the disease can be contracted via needles, barber's instruments, an infected butcher, the menstrual blood of an infected woman or the urine of an infected person;

Category 4: those trying to find out more about the symptoms, how they can be recognized, the incubation period of the virus, and the possibility of a cure;

Category 5: enquiries as to whether AIDS can be insect borne;

Category 6: those enquiring how they can discover whether their intended spouse is 'clean', how they can protect themselves, and how to protect those who are just beginning their sexual lives;

Category 7: those asking that more radio time should be devoted to the disease and wanting more hard information about the rates of infection in the population;

Category 8: those asking for educational material;

Category 9: those describing their own symptoms and/or those of their friends and relatives and asking for specific advice.

Partial analysis of these letters indicates that there was a large demand for information. The consistently highest number of letters fell into Category

4 – enquiries for information – and Category 8 – those asking for information to be sent. Most reassuring, in the light of what has been said about rationalizations that deflect real concern from the issues, is the small numbers of letter falling into Category 1. However, such a conclusion cannot be generalized to the whole population, given that the responses were from those who both had access to a radio, listened to it and were prepared and able to write a letter – undoubtedly the more sophisticated sections of the population (Tuju, February 1990, p. 5).

In Zaïre, attempts have been made to encourage condom use by employing high-pressure marketing techniques. It is reported that a programme sponsored by the United States Population Services International had increased the numbers of condoms sold from 200,000 in 1986 to around nine million in 1990 (*WorldAIDS*, January 1991, p. 3). The methods used included price subsidies and reduced entry to pop concerts for those carrying a box of Prudence condoms. However, these figures do not indicate whether or not the condoms sold are used, or whether they are used correctly. Indeed, the use of sales techniques such as those reported here are no confirmation of success unless follow-up studies are carried out to discover whether these are just sales or uses: research in many countries has indicated deep-seated resistance to their use. In the Central African Republic, for example, even among those who had received information about the disease, most saw condoms negatively and only 15 per cent of an urban sample actually claimed to use them (Gresenguet et al., 1989, pp. 48–53). Here is more evidence of the difficulties of changing attitudes and behaviour and an indication that more sophisticated techniques of market research and marketing are needed.

In Zambia, serious attempts have been made to use multi-media outlets to get the message over to the population of the Copperbelt – the location of 45 per cent of recorded AIDS cases, but home to only 25 per cent of the population. Large employers, such as Lonrho, certainly take the issue seriously, introducing AIDS education for the workforce and distributing some condoms (private communication). Under the auspices of the Copperbelt Health Education Project (CHEP) information about AIDS is serialized once a week on the front pages of two national daily newspapers published in English. In addition, the radio carries a weekly programme 'Health is Wealth' in six national languages and sheet metal hoardings by roadsides and smaller notices attached to litter bins provide information about the disease. In an effort to get the message over to the school-age children, slogans have been printed on the covers of four million exercise books distributed to schools throughout the country. Efforts to contact the wider urban population have included pamphlets sent out with water bills. Traditional theatre and popular music have also been used (Mouli, September 1989, p. 4). However, it must be noted, most of these efforts are aimed at the urban population and, as we have suggested, a major issue in Africa is bringing the message home to rural populations, few of whom see a newspaper or listen to the radio. In neighbouring Zimbabwe, the Harare Women's Action Group has published a cartoon book, *AIDS: Let*

Us Fight it Together in Shona and Ndebele in an attempt to get educational messages over to ordinary people in language they can understand (Rider, 1989, p. 5); in South Africa, puppet theatre has been used to communicate messages about safe sex and condom use (*WorldAIDS*, No. 4, July 1989, p. 5).

All of the foregoing brings us to the question of what can and should be done. It has been commented that: 'If we could play at being Satan for the day, charged with the task of designing an epidemic to undermine both the developed and underdeveloped countries of the world at the end of the twentieth century, then the blueprint for the design would incorporate many of the features of AIDS' (Connor and Kingman, p. 1, 1988). As the reader will have seen from this book, the results are terrible, at the personal level most immediately and perhaps above all, but also, as we have argued throughout, potentially at the level of whole societies in Africa and possibly elsewhere. One of the ways in which the disease is most terrifying is that it undermines political and social order in a peculiar way. Governments, whatever their complexion and source of legitimacy (and it has to be remarked that many of the governments of affected countries in Africa have precious little in the way of legitimacy), claim to govern by virtue of what they can offer their people. In the introduction to this book, we argued that in Africa, the AIDS pandemic confronts us with the full range of development issues in a particularly acute form. It raises one specific issue, which is not restricted to the countries of Africa, although of major importance there, given the political fragility of many governments. This is the issue of what governments do and say when they are confronted with a problem they cannot 'solve', for which they can offer little even in the 'longer' political term (probably about five years), which threatens that elusive goal 'development', so often their rationale and with which they cannot effectively cope.

This question was put in another form to a meeting of social scientists at Lake Kariba in Zambia in 1990, when Dr Beth Maina-Ahlberg, a Kenyan sociologist, said: 'We have knowledge, but what do we do with it? Doctors take blood and go away, social scientists ask questions and go away. But the community is still dying' (Bond, 1990, p. 5). As we have seen, government responses to the pandemic have been patchy, constrained by lack of resources, the need to balance political, economic, religious and social interests and forces in each society, and in some cases the delay in facing up to the implications of the disease. Thus, as we argued in the introduction, AIDS raises the whole range of development problems.

The precise impact AIDS will have on the different countries of Africa will depend upon the specific features of each one and on its own particular economic, political and cultural circumstances. In this case-study of Uganda we have shown how the epidemic has affected one country, and have hazarded some indications of the expected longer term impacts. More or less widespread changes in the dependency ratio, declines in food production and shortages of skilled personnel are all possible foreseen effects. The impacts of structural-adjustment policies, the mismanagement of economic policy and in some cases massive population movements and increasing problems of ungovernability and civil disorder will all contribute to the difficulties of tackling this pandemic. Any realistic response requires effective administration and governance for the effective provision of services to the affected populations. The international response so far through the World Health Organization and its General Programme on AIDS has at least established a basis for national monitoring of the magnitude of the problem in most countries. However, this is really only a beginning. The quality of the information produced by the different national AIDS programmes is very variable. But a longer term perspective, one that will deal with a long wave crisis such as we have described in this book, requires much more than this basic effort. It requires above all a degree of openness about what is happening.

Uganda is an example of where such frankness about the problem, despite very understandable local sensitivities, at least means that the scale of the problem is widely appreciated within the country and thus by those who are trying to confront it. But even given such openness, what can really be done? One response is to appreciate the strength of what we have been saying throughout this book – that the longer term impact of the disease in any African country will only be a reflection of the much broader economic, social and political problems of some of the world's poorest countries. Thus the response to the disease must take into account the context in which it is occurring. The countries of Africa do not, on the whole, have effective administrations. Their governments often lack legitimacy. They are deeply indebted. The position of women is one of great vulnerability. And yet, for all this, as we have seen, in Uganda, people are coping as they have always coped. But we know that secular declines in health, production and social order can and do result in people having to cope at a lower and lower level. And in the end, individuals die, communities decline and even whole societies reach a situation of what looks like terminal decay – as in Somalia and the Sudan. AIDS may be just another turn of the screw in this downward spiral for many countries. Yet people go on coping. We can learn from their experience and their inventiveness, and perhaps, just perhaps, something can be done through the actions of non-governmental organizations working with local communities in ways that will enable them to avoid the worst long-term effects of this pandemic.

The VIIth International AIDS Conference in Florence in June 1991 clearly revealed two features that have been apparent for some time to those working in the field of the downstream impact of the AIDS epidemic. These two features are concerned with two great divisions: the division between medical and social scientists and that between treatment, care and concern for the wealthy and the opposites for the poor. Each of these divisions is to be seen on a local, national and, perhaps most worryingly of all, an international scale.

In this book, we have discussed some of the downstream issues raised by the AIDS pandemic. These are issues of social and economic life. To state the obvious, AIDS affects people. However, most of the resources are being put into sophisticated medical research. This is quite right: the search for a cure or a vaccine must take the highest priority - but not to the total exclusion of funding for projects aimed at confronting the social and economic impacts. In the absence of either a vaccine or a cure, and given the long wave nature of the crisis, we must begin to take seriously the problems of long-term responses to its social and economic effects. The kinds of medical responses that are likely are very expensive. Some suggested treatments that might keep a person with AIDS alive for 20 years will cost in the region of $85,000 per person. Such sums are an enormous burden even for the health budgets of very rich countries; they are impossible for all others.

Africa, and most other parts of the world, are going to be terribly affected. The falling price of commodities to pay for even the basic medicines and testing equipment and the absence of other infrastructure to prevent further spread, all make Africa particularly vulnerable. In the meantime, an undercurrent of opinion is beginning to suggest that AIDS is under control in Europe and North America, that it can now be seen as 'just' another tropical disease - like malaria - against which the people of Europe and North America can protect themselves by means of simple precautionary measures. Such attitudes are easy to adopt, they fit well with established prejudices along class, gender and ethnic lines. They insidiously penetrate research agendas. For such reasons, those of us concerned about confronting the social and economic impact of this pandemic face a considerable struggle to have the problem recognized as a legitimate one for funding. As Jonathan Mann said recently:

> The pandemic not only remains dynamic, volatile and unstable, but it is gaining momentum - and its major impacts, in all countries, are yet to come. Public complacency is rising and societal commitment against HIV and AIDS is declining. (The *Guardian*, 26 June 1991)

We believe it is essential to shake that public complacency about the situation within Africa and towards Africa. In the end, the issue comes down to the capability of the international community to act as a community and to mobilize in response to the pandemic in Africa. At present, the international community is not coping; it is experimenting rather too slowly as the crest of this long wave disaster begins to break. In the end, it is the people themselves who have been suffering the disease and who are shouldering its impact in caring, looking after orphans and working harder in the fields who have to cope, to adapt and to endure. The trouble is that they cope with the downstream effects more or less well, maybe without too many consequent disasters, but they, like anyone, anywhere, are not really able to cope with the upstream effects - i.e. avoid infection - without outside assistance. As the pandemic bites further into the lives and livelihoods of African people it will gradually erode their ability to cope and the coping mechanisms which they have developed. Outside assistance is traditionally, in the idiom of the Western nation state, the role of good government. And good government, in many parts of Africa, is missing. So we come full circle, back to Nankya, whom we met in Chapter 1. She and her children still await assistance and support as they face the AIDS pandemic in Africa.

References

Abbas, N., Muller, D., 'The Impact of AIDS Mortality on Children's Education in Kampala (Uganda)', *AIDS Care*, (1990), Vol. 2, No. 1, pp. 77–80.

Abel, N., Barnett, T., Blaikie, P. M., Cross, J. S. W., Bell, S., 'The Impact of AIDS on Food Production Systems in East and Central Africa over the next ten years: a programmatic paper', in A. F. Fleming et al. (eds), *The Global Impact of AIDS*, Allan R. Liss, New York, and John Wiley, London, 1988), pp. 145–54.

Africa Analysis, (2 February 1990), p. 2.

Africa Analysis, 'Zimbabwe admits AIDS increase', (17 May 1991), p. 14.

AIDS Analysis Africa, Southern Africa Edition, Vol. 1, No. 1, (June/July1990), p. 9.

AIDS Analysis Africa, Southern Africa Edition, Vol. 1, No. 5 (Feb/March 1991) p. 11.

AIDS Analysis Africa, Vol. 1, No. 1, (March–April, 1991), p. 5.

AIDS Analysis Africa, 'Botswana votes funding: but no vending machines', Vol.1, No. 2, (May/June 1991), p.6.

AIDS Analysis Africa, Vol. 1, No. 2, (May/June 1991), pp. 6–7.

AIDS Analysis Africa, Vol. 1, No. 2 (May/June 1991), p.1.

AIDS Newsletter, Vol. 4, Issue 13, (October 1989), Item 729, p. 2.

Amat, J. M. et al., 'Apports de la Géographie à la Compréhension de l'épidemiologie des virus VIH dans l'Afrique sud-Saharienne', Poster 056 at the IVth International Conference on AIDS and Associated Cancers in Africa, Marseille, France, 18–20 October 1989.

Anderson, R. M., 'Editorial Review: Mathematical and Statistical Studies of the Epidemiology of HIV', *AIDS*, (1 June 1989), Vol. 3, No. 3, pp. 333–46.

Anderson, R. M., May, R. M., McLean, A. R., 'Possible Demographic Consequences of AIDS in Developing Countries', *Nature*, Vol. 332, (1988), pp. 228–34.

Anderson, R. M., Boily, M. C., May, R. M., Wan Ng, T., 'The Influence of Differential Sexual-Contact Patterns between Age and Classes on the Predicted Demographic Impact of AIDS in Developing Countries', in *Biomedical Science and The Third World: Under the Volcano*, Volume 569 of the *Annals of the New York Academy of Sciences*, (8 December, 1989), pp. 240–74.

Anderson, R. M., May, R. M., Boily, M. C., Garnett, G. P., Rowley, J. T., 'The spread of HIV-1 in Africa: sexual contact patterns and the predicted demographic impact of AIDS, *Nature*, Vol. 352, (15 August, 1991), pp. 581–9.

Ankrah, E. M., 'AIDS: Methodological Problems in Studying its Prevention and Spread', *Social Science and Medicine*, (1989), Vol. 29, No. 3, pp. 265–76.

Anonymous (1), 'AIDS in Africa', *AIDS – Forschung*, (1987), Vol. 2, pp. 5–25.

Anonymous (2), 'The AIDS Crisis in Anglophone Africa: An Early Assessment of the Demographic and Socioeconomic Implications', Canadian International Development Agency, Anglophone Africa Branch, (July 1987).

Apted, F. C., 'The Epidemiology of Rhodesian Sleeping Sickness', in H. W. Mulligan (ed.), *The African Trypanosomiases*, (London, Allen and Unwin, 1970), pp. 645–60.

Asedri, V., '790,522 Ugandans are HIV Positive', *The New Vision*, (Kampala, 1989), Vol. 4, No. 271, p.1.

Baker, K., *WorldAIDS*, (November 1989), No. 6, p. viii.

Barnett, T., *The Gezira Scheme: an illusion of development*, (Frank Cass, London, 1977).

Barnett, T., Blaikie, P. M., 'The Possible Impact of AIDS upon Food Production in East and Central Africa', *Food Policy*, (February 1989), Vol. 14, No. 1, pp. 2–6.

Bassett, M. B., Mhloyi, M., 'Women and AIDS in Zimbabwe: The Making of an Epidemic', *International Journal of Health Services*, (1991) Vol. 2, No. 1, (1991), pp. 143–56.

Bean, J. M. W., 'The Black Death: the Crisis and its Social and Economic Consequences' in D. Williman (ed.), *The Black Death: The Impact of the Fourteenth Century Plague*. Center for Medieval and Early Renaissance Studies, Binghamton, New York. (1982), pp. 23–38.

Beer, C., Rose, A., Tout, K., 'AIDS – The grandmother's burden', in *The Global Impact of AIDS*, in A. F. Fleming et al. (eds), (Alan R. Liss, New York, 1988), pp. 171–82.

Bellaby, P., 'To risk or not to risk? The grid/group approach, risk cultures and hazards to life and limb', unpublished paper delivered to the British Sociological Association Medical Sociology Group Annual Conference, Manchester Polytechnic, 1990.

Benslimane, A., Riyad, M., Sekkat, S. et al., 'Incidence of HIV infections in Morocco, Second International Symposium on AIDS and Associated Cancers in Africa, Abstract TH-84, (Naples, 1987), p. 114.

Berkley, S. F., Okware, S., Naamara, W., 'Surveillance for AIDS in Uganda', *AIDS*, (1988), Vol. 3, No. 2, pp. 79–85.

Berkley, S. F., Naamara, W., Okware, S., Downing, R., Konde-Lule, J., 'The Epidemiology of AIDS and HIV infection in Women in Uganda', poster 012 at the IVth International Conference on AIDS and Associated Cancers in Africa, (18–20 October 1989), Marseille, France.

Berkley, S. F., Downing, R., Konde-Lule, J. K., 'Knowledge, Attitudes and Practices Concerning AIDS in Ugandans', *AIDS*, Vol. 3, (1989), pp. 513–18.

Berkley, S. F., Naamara, W., Okware, S. et al., 'AIDS and HIV infection in Uganda – are more women infected than men?', *AIDS*, (1990), Vol. 4, pp. 1237–42.

Biberfeld, G., et al., 'Prevalence of HIV-1 Infection in the Kagera Region of Tanzania: a population based study', *AIDS Journal*, (1990), Vol. 4, pp. 1081–5.

Biritwum, R. B. et al., 'Mechanisms for Research Collaboration: Analysis of AIDS Research Inventory in Africa', p. 163 in the Abstract Volume of the IVth International Conference on AIDS and Associated Cancers in Africa, (1989), Marseille.

Blaikie, P. M., Cameron, J. D., Fleming, R., Seddon, J. D., 'Centre, Periphery and Access: Social and Spatial Relations of Inequality in West-Central Nepal', Development Studies Monograph No. 5, (1977), School of Development Studies, University of East Anglia, Norwich NR4 7TJ, UK.

Blaikie, P. M., *The Political Economy of Soil Erosion in Developing Countries*, (Longman, London and New York, 1985).

Bond, G. C., Vincent, J., 'Living on the edge of structural adjustment in the context of AIDS', in H-B Hansen and M. Twaddle (eds), *Uganda: Structural Adjustment and Change*, (James Currey, London, 1990).

Bond, V., 'A Question of Sensitivity', *WorldAIDS*, (July 1990), No. 10, p. 5.
Bongaarts, J., 'Modelling the Demographic Impact of AIDS in Africa', *AAS Symposia Paper*, R. Kulstad, (ed.), *AIDS*, (1988).
Bongaarts, J., Way, P., 'Geographical Variation in the HIV Epidemic and the Mortality Impact of AIDS in Africa', Working Paper 1, (1989), Population Council, Centre for Policy Studies.
Boserup, E., *Woman's Role in Economic Development*, (Allen & Unwin, London, 1970).
Boserup, E., 'Economic and Demographic Inter-relationships in Sub-Saharan Africa', *Population and Development Review*, (1985), Vol. 11.
Brokensha, D., MacQueen, K., Stess, L., 'Anthropological Perspectives on AIDS in Africa - Priorities for intervention and research', AIDSTECH Project, (1988), USA.
Brown, P., 'Africa's Growing AIDS Crisis', *New Scientist*, (17 November 1990), p. 41.
Brown, P., 'Why straight sex is not safe sex', *New Scientist*, (27 July 1991), No. 1779, pp.17–18.
Brown, P., 'Zaïre Tests Spermicides as Prevention against HIV', *AIDS Analysis Africa*, (May–June 1991), Vol.1 No. 2, p. 5.
Brown, P., Concar, D., 'HIV epidemic threatens Asia's developing nations', *New Scientist*, (22 June 1991), p. 18.
Bukedde, (September 11 1989), Vol.1, No. 3.
Caldwell, J. C, Reddy, P. H Caldwell, P., 'Periodic High Risk as a Cause of Fertility Decline in Changing Rural Environment: Survival Strategies in the 1980–1983 South Indian Drought', *Economic Development and Cultural Change*, (1986), pp. 677–701.
Caldwell, J. C, Caldwell, P., Quiggin, P., 'The Social Context of AIDS in sub-Saharan Africa', *Population and Development Review*, (1989), Vol. 15, No. 2, pp. 185–234.
Campbell, I., Williams, G., *AIDS Management: an integrated approach*, (Actionaid, Hamlyn House, Archway, London in association with AMREF and World in Need, 1990).
Carael, M. et al., 'Socio-Cultural Factors In Relation to HTLV-III/LAV Transmission in Urban Areas in Central Africa', International Symposium on African AIDS, (Brussels, 22–23 November 1985), Abstract 05.I.
Carvalho, N., Mawaali, W., Moser, R., Carswell, J. W., 'Can simple Demographic Data identify Blood Donors at 'Very Low' risk of HIV infection in Urban Africa?', Abstract No. 385, p. 199 in the abstract volume to the 6th International Conference on AIDS and Associated Cancers in Africa, (Marseille, 1989).
Chin, J., Lwanga, S., Mann, J., *The Global Epidemiology and projected short-term demographic impact of AIDS*, mimeo, (August 1988), World Health Organization, Geneva.
Chin, J., Mann, J., 'Global Surveillance and forecasting of AIDS', *Bulletin of the World Health Organisation*, (1989) Vol. 67, No. 1, pp. 1–7.
Chin, J., Sato, P., Mann, J., *'Estimates and Projections of HIV/AIDS To the Year 2000'*, mimeo, (1989), World Health Organization, Geneva.
Chin. J., 'Current and future dimensions of the HIV/AIDS pandemic in women and children', *The Lancet*, (1990), No. 336, pp. 221–4.
Chirimuuta, R. C, Chirimuuta, R. J, *AIDS, Africa and Racism*, (Bretby House, Stanhope, Derbyshire DE15 OPT, UK, 1987).
Christenson, B. A., 'Las implicaciones del SIDA en la fuerza de trabajo en México', *Salud Pública de México*, (1988) Vol. 30, No. 4, pp. 619–28.

Cleave, J. H., *African Farmers: Labor Use In The Development of Smallholder Agriculture*, (Praeger, London, 1974).
Colbourne, M., *Malaria in Africa*, (Oxford University Press, London, 1966).
Collinson, M. P., *Farming systems Research in Eastern Africa: The experience of CIMMYT and some National Agricultural Research Services*, (1976–81), Michigan State University International Development Paper No. 3, (1982), Department of Agricultural Economics, Michigan State University, East Lansing.
Conant, F. P., 'Social Consequences of AIDS: Implications for East Africa and the Eastern United States', *AIDS 1988: AAAS Symposia Paper*, R. Kulstad (ed.), (1988).
Connor, S., Kingman, S., *The Search for the Virus: the scientific discovery of AIDS and the quest for a cure*, (Penguin Books, London, 1988).
Corbett, J., 'Famine and Household Coping Strategies', *World Development*, (1988), Vol. 16, No. 9, pp. 1099–112.
Courier, The, 'AIDS: The Big Threat', (March–April 1991), No. 126, pp. 42–76.
Croze, H., *GRID: Uganda Case Study: A Sampler Atlas of Environmental Resource Databases within GRID*, (Nairobi, 1987).
Cutler, P., 'The Response to drought of Beja famine refugees in Sudan', *Disasters*, (10 March 1986), pp. 181–8.
Das Gupta, M., 'Informal Security Mechanisms and Population Retention in Rural India', *Economic Development and Cultural Change*, (1986), pp. 101–20.
Davenport–Hines, R., *Sex, Death and Punishment: attitudes to sex and sexuality in Britain since the Renaissance*, (Collins, London, 1990).
Douglas, M., *Purity and Danger*, (Penguin, Harmondsworth, 1966).
Douglas, M., *Risk Acceptability According to the Social Sciences*, (Russel Sage Foundation, New York, 1985).
Douglas, M., Wildavsky, A., *Risk and Culture: an essay on the Selection of Technological and Environmental Dangers*, (University of California Press, Berkeley, 1982).
Dunn, A., Hunter, S., Nabongo, C., Ssekiwanuka, *Enumeration and Needs Assessment of Orphans in Uganda: Survey Report*, Social Work Department, Save the Children, POB 1124, Kampala, Uganda, April, 1991.
Dyakire, A., 'Museveni asks for legislation to protect minors', *WorldAIDS*, (May 1989), No. 3, p. 4.
Economic and Social Council, United Nations, Global Strategy for the Prevention and Control of Acquired Immunodeficiency Syndrome (AIDS), (26 May 1989, United Nations, New York).
Feurstein, M., Lovel, H., 'Seeing Light at the End of the Tunnel', *Community Development Journal*, Vol. 24, No. 3, (July 1989). *Financial Gazette*, Harare, (6 July 1990).
Finney–Hayward, R., 'Orphans and AIDS', *AIDS and Society*, Vol. 1, No. 4 (July 1990), p. 8.
Fitzgerald, M. A., 'AIDS Cuts Deadly Swathe Through African Leaders', *The Sunday Times*, (1 July 1990), London.
Fleming, A. F., 'Seroepidemiology of human immunodeficiency viruses in Africa', *Biomedicine and Pharmacotherapy*, (1988a), Vol. 42, pp. 309–20.
Fleming, A. F., 'AIDS in Africa – an update', *AIDS-Forschung*, (1988b), Vol. 3, pp. 116–38.
Fleming, A. F., 'AIDS in Africa', *Balillière's Clinical Haematology*, (January 1990), Vol. 3, No. 1, pp. 177–205.
Fleming, A. F., 'Lessons from Tropical Africa for South Africa in addressing the impact of the HIV and AIDS epidemic', unpublished paper delivered to the Workshop to Assess the Economic Impact of AIDS in South Africa, University

of Natal, Durban, (22–23 July, 1991).
Fleming, A. F., Carballo, M., FitzSimons, D.W., Bailey, M.R., Mann, J., *The Global Impact of AIDS*, (Alan R. Liss, Inc., New York, 1988).M
Ford, J., *The role of the trypanosomiases in African ecology: a study of the tsetse-fly problem*, (Oxford University Press, Oxford, 1971).
Forster, S. J., Furley, K. E., '1988 Public Awareness Survey on AIDS and Condoms in Uganda', *AIDS*, (1989), Vol. 3, No. 3, pp. 147–54.
Gallo, R., *Virus Hunting: AIDS, Cancer, and the Human Retrovirus: a story of Scientific Discovery*, (New Republic/Basic Books, New York, 1991).
Garine, de, I., Harrison, G. A., *Coping with uncertainty in food supply*, (Clarendon Press, Oxford, 1988).
Giller, J. E., Bracken, P. J., Kabaganda, S., 'Uganda, War, Women and Rape', *The Lancet*, (1991), No. 337, p. 604.
Gillespie, S., 'The Potential Impact of AIDS on Food Production Systems in Central Africa', (unpublished report, Food and Agricultural Organization of the United Nations, Rome, 1989a).
Gillespie, S., 'Potential Impact of AIDS on Farming Systems: a case-study from Rwanda', *Land Use Policy*, (1989b), Vol. 6, pp. 301–12.
Gluckman, M., *Custom and Conflict in Africa*, (Basil Blackwell, Oxford, 1955).
Goodgame, R. W., 'AIDS in Uganda: Clinical and Social Features', *New England Journal of Medicine*, (1990), Vol. 323, No. 6, pp. 383–9.
Gottfried R. S., *The Black Death*, (The Free Press, New York, 1983).
Government of Uganda, *Decree on Land*, Uganda Government Printers, (Entebbe, 1975).
Government of Uganda, Ministry of Health, *Demographic Health Survey, Preliminary Report 1988–89*, Entebbe/Demographic Health Surveys, (Institute of Resources Development/Westinghouse, 1989).
Government of Uganda, Ministry of Health, AIDS Control Programme, Press Release of the National Serosurvey for Human Immunodeficiency Virus (HIV) in Uganda, Entebbe, 1 December 1989.
Government of Uganda, Ministry of Health, AIDS Control Programme, *Monthly Report*, (Entebbe, 1989).
Government of Uganda, Ministry of Health, AIDS Control Programme, *Quarterly AIDS Surveillance Report*, (Entebbe, 1990).
Green, R. H., 'Magendo in the political economy of Uganda: pathology, parallel system or dominant sub-mode of production?', Institute of Development Studies, University of Sussex, *Discussion Paper No. 164*, (1981).
Gregson, S., 'Predicting the Future Impact of AIDS in Uganda', unpublished MSc Dissertation, London School of Economics, (1990).
Gresenguet, G. et al., 'Connaisance, attitudes et croyance sur le SIDA', *Médecine d'Afrique Noire*, (1989), Vol. 36, No. 1, pp. 48–53.
Guardian, The, London and Manchester, (26 April 1991).
Guardian, The, London and Manchester, (17 June 1991).
Guardian, The, London and Manchester, (18 June 1991).
Guardian, The, London and Manchester, (26 June 1991).
Hampton, J., *Living Positively with AIDS: The AIDS Support Organization (TASO), Uganda*, Actionaid, (Hamlyn House, Archway, London in association with AMREF and World in Need, 1990).
Hansen, H. B., Twaddle, M. (eds), *Uganda Now: Between Decay and Development*, (James Currey, London, 1988).
Hansen, H. B., Twaddle, M. (eds), *Uganda: Structural Adjustment and Change*, (James Currey, London, 1990).

Hartwig, G. W., Patterson, K. D., *Diseases in African History: an Introductory survey and case studies*, (Duke University Press, Durham, N. Carolina, 1978).
Hatcher, J., *Plague, Population and the English Economy 1348–1530*, (Macmillan, London, 1977).
Health Education Authority, *Operational Plan 1991–93*, (Health Education Authority, London, 1991a).
Health Education Authority, *Health Update No. 2: Smoking*, (Health Education Authority, London, July, 1991b).
Hecker, J. F. C., *The Epidemics of the Middle Ages*, (3rd edn.), Translated by B. G. Babington, (Trubner, London, 1859).
Holmes, W., 'Condom use – the catch 22', *WorldAIDS*, (January 1991), No. 13, p. 9.
Hooper, E., *Slim: A Reporter's Own Story of AIDS in East Africa*, (Bodley Head, London, 1990).
Hunter, S., untitled paper given at the NGO AIDS Coordinating Committee Conference, 1 December 1989, Institute of Education, University of London, (1989a).
Hunter, S., 'Social Aspects of AIDS in Uganda in Perspective', unpublished paper, (1989b).
Hunter, S., 'AIDS: Community Coping Mechanisms in the face of exceptional demographic change', an update on the orphan situation in Uganda, paper given to the International Conference Centre, Kampala, (20 November 1990a).
Hunter, S., 'Orphans as a window on the AIDS epidemic in sub-Saharan Africa: initial results and implications of a study in Uganda', *Social Science and Medicine*, (1990b), Vol. 31, No. 6, pp. 681–90.
Hunter, S., Dunn, A., 'Enumeration and Needs Assessment of children orphaned by AIDS in Uganda', poster 292 at the IVth International Conference on AIDS and Associated Cancers in Africa, (18–20 October 1989), Marseille, France, (1989).
Hunter, S., Seronjogi, L., Barton, T., *Proposal to Investigate The Nutritional and Health Status of Children in Special Circumstances in Masaka, Rakai and Kalangala Districts*, Uganda, UNICEF, (Kampala, 1990).
International Bank for Reconstruction and Development (World Bank), *World Development Report*, (World Bank, Washington, 1988).
International Bank for Reconstruction and Development (World Bank), *World Development Report*, (World Bank, Washington, 1989).
Johnson, D. T., Ssekitoleko, Q. W., *Current and Proposed Farming Systems in Uganda*, Farm Management and Economic Research Station, Planning Division, (Ministry of Agriculture, Entebbe, Uganda, 1989).
Kaijuka, E. M., Kaija, E. Z. A, Cross, A. R., Loaiza, E, *Uganda: Demographic and Health Survey 1988–89*, (Ministry of Health, Government of Uganda, Uganda, October 1989).
Kalambay, K. et al., 'LAV/HTLV-III seroprevalence among patients without AIDS or AIDS related complex hospitalised at the University Hospital, Kinshasa, Zaïre', Poster 371, International Conference on AIDS, Paris, (23–25 June 1986), Abstracts, p. 129.
Kaleeba, N., Kalibala, S., 'AIDS and Community-based Care in Uganda: The AIDS Support Organization, TASO', *AIDS Care*, (1989), Vol. 1, No. 2, pp. 173–5.
Kandyoti, D., 'Bargaining with Patriarchy', *Gender and Society*, (1988), Vol. 2, No. 3, pp. 274–90.
Kisekka, M. N., Otesanya, B., 'Sexually Transmitted Diseases as a Gender Issue: Examples from Nigeria and Uganda', Paper given at the AFARD/AAWORD

Third General Assembly and Seminar on The African Crisis and the Women's Vision of the Way Out, Dakar, Senegal, (August, 1988).
Kjekshus, H, *Ecology Control and Economic Development in East African History: the case of Tanganyika 1850–1950*, (Heinemann, London, 1977).
Konde-Lule, J. K., 'Group Health Education Against AIDS in Rural Uganda', *World Health Forum*, (1988), Vol. 9, No. 3, p. 384.
Kouassi, M., 'Infection Rises in Ivory Coast', *WorldAIDS*, (May 1989), No. 3, p.4.
Kyewalyanga, F-X. S., *Traditional Religion, Custom and Christianity in Uganda*, privately published at Freiburg im Breisgau, (1976), printed by Offsetdrückerei Johannes Krause, Freiburg i. Br.
Kyeyune-Ssentongo, L. L., *An Economic Study of the Farming System of Central and South Ankole*, Government of Uganda, Ministry of Agriculture and Forestry, Entebbe, (Makerere Insititute of Social Research, Kampala, 1985).
Larson, A., 'The Social Context of HIV Transmission in Africa: A review of the historical and cultural bases of East and Central African sexual relations', *Working Paper No.1*, (The Australian National University, Canberra, 1989).
Lavreys, L. et al., *Retroviral Infection in adults at Dabou Protestant Hospital, Ivory Coast*, Paper 042 at the IVth International Conference on AIDS and Associated Cancers in Africa, (18–20 October 1989), Marseille, France.
Lesotho Today, Maseru, (6 December 1990), quoted in *AIDS Analysis Africa*, Southern Africa Edition, (February–March 1991), Vol. 1, No. 5, p. 11.
Lewis, P., 'AIDS and AIDS in Africa', unpublished mimeo, Romneya, St Chad's Avenue, Midsomer Norton, Bath, UK, (April 1991).
Longhurst, R., 'Household Food Strategies in Response to Seasonality and Famine', *IDS Bulletin*, (1986), Vol. 17, No. 3, pp. 27–35.
Lubega, J., Muller, O., Senoga, J., 'The Impact of the AIDS Education Programme on Ugandan Schoolchildren', *AIDS Care*, (1989), Vol. 1, No. 2, pp. 135–6.
Lutunda, S., Kabwe, C., 'Church and State at odds in Zambia', *WorldAIDS*, (March 1989), No. 2, p. 5.
Lyerly, W., Barry, S., Miller, N., 'HIV Transmission and the Military', *AIDS and Society: International Research and Policy Bulletin*, (July 1990), Vol. 1, No. 4, pp. 1 and 8–10.
Lyons, M., 'Sleeping Sickness, colonial medicine and imperialism: some connections in the Belgian Congo', in R. M. Macleod and M. Lewis, *Disease, Medicine and Empire*, (Routledge, London, 1988a).
Lyons, M., 'African Trypanosomiasis' (sleeping sickness), paper delivered at the Institute of Commonwealth Studies and the London School of Hygiene and Tropical Medicine, (1988b).
Maclean, G., 'The Relationship Between Economic Development and Rhodesian Sleeping Sickness in Tanganyika Territory', *Annals of Tropical Medicine and Parasitology*, (1929), Vol. XXIII, pp. 37–46.
Macleod, R. M., Lewis, M., *Disease, Medicine and Empire*, (Routledge, London, 1988).
Mamdani, M., 'Class Struggles in Uganda', *Review of African Political Economy*, (November 1975), No. 4, pp. 26–61.
Mamdani, M., 'NRA/NRM: two years in power', Public lecture, Makerere University, (3 March 1988), Published in *Weekly Topic*, (16, 23 and 30 March 1988, Progressive Publishing House, Kampala).
Mann, J., 'Worldwide epidemiology of AIDS', in Fleming et al., *The Global Impact of AIDS*, (Alan R. Liss, New York, 1988), pp. 3–7.
McMaster, D. N., *A Subsistence Crop Geography of Uganda*, (Geographical Publications Ltd., Kampala?, 1962).

McNeill W. H., *Plagues and Peoples*, (Basil Blackwell, Oxford, 1976).

Merson, M., 'We have to think of AIDS as a development problem not just a health one', interview in *The Courier*, (March–April, 1991), No. 126, pp. 50–2.

Mmbaga, C., Ogola, H., '"Let it kill me" say the men', *WorldAids*, (March 1989), No. 2, p. 3.

Morbidity and Mortality Weekly Report, (5 June 1981), No. 30, pp. 250–2.

Mouli, C., 'Using Metal Sheets and Disco Beats', *WorldAIDS*, (September 1989), No. 5, p. 4.

Mukoyogo, M. C., Williams, G., *AIDS Orphans: a community perspective from Tanzania*, (ActionAid, Hamlyn House, Archaway, London N19 5PG, UK; AMREF, Wilson Airport, POB 30125, Nairobi, Kenya; AMREF, POB 2773, Dar es Salaam, Tanzania; World in Need, 42, Culver St East, Colchester, CO1 1LE, UK, 1991).

Mukwaya, A. W., *The Busulu Envujjo Law*, (East African Institute for Social Research, Makerere, 1954).

Munyako, D., 'Rural realities and condom use', *WorldAIDS*, (May 1991), No. 15, p. 3.

Murphy, L. M., Moriarty, A. B., *Vulnerability, Coping and Growth*, (Yale University Press, New Haven and London, 1976).

Musagara, M., Musgrave, S., Biryahwaho, B., Serwadda, D., Wawer, M., Konde-Lule, J., Berkley, S., Okware, S., 'Sero-prevalence of HIV-1 in Rakai District, Uganda', Poster 010 at the IVth International Conference on AIDS and Associated Cancers in Africa, (18–20 October, 1989), Marseille, France.

Musoke, D., Tamalie, N., 'Songs to persuade the sceptics', *WorldAIDS*, (January 1990), No. 7, p. 8.

Nabarro, D., McDonell, C., 'The Impact of AIDS on Socioeconomic Development', *AIDS*, (Supplement 1, 1989), Vol. 3, pp. S265–S272.

Nalwanga-Sebina, A. J., 'Survey on coping mechanisms in circumstances of exceptional demographic change: Rakai and Kabale Districts of Uganda', unpublished report, Makerere Institute of Social Research, Kampala, (July 1989).

New Vision, Kampala, Uganda, (1 December 1989).

Newman F. X., (ed.), *Social Unrest in the Late Middle Ages*, (Binghamton, New York, 1986).

Norse, D., 'Food Supply in sub-Saharan Africa', *AIDS Newsletter*, Vol. 6, Issue 11, August, 1991, pp. 11–12.

Obbo, C., 'Is AIDS Just Another Disease?', *AAAS Symposia Paper*, (1988).

ODNRI (Overseas Development Natural Resources Institute), 'Potential Impact of AIDS on Food Production and Consumption: Tabora Case Study', for the Overseas Development Administration, Eland House, Stag Place, London, Contract No. C 0863, (1989).

Ogola, H., 'Unequal under Tanzanian Law', *WorldAIDS*, (January 1991), No. 13, p. 5.

Ogunseitan, O., 'Striking the right chord', *WorldAIDS*, (November 1989), No. 6, p. 5.

Okware, S. I., 'Towards a National AIDS-Control Program in Uganda', *The Western Journal of Medicine*, (1987), Vol. 146, Part 6, pp. 726–9.

Over, M., 'The Economic Impact of Fatal Adult Illness from AIDS and other causes in Sub-Saharan Africa: A Research Proposal', unpublished document, Research Department, World Bank, (Washington, 1990).

Over, M., et al., 'The Direct and Indirect Costs of HIV infection in developing countries: the cases of Zaïre and Tanzania', in A. R. Fleming et al. (eds), *The Global Impact of AIDS*, (Alan R. Liss, New York, 1988), pp. 123–35.

Packard, R. M., *White Plague, Black Labour: Tuberculosis and the political economy of health and disease in South Africa*, (University of Natal Press, Pietermaritzburg and James Currey, London, 1989).

Parker, M., 'South Africa – Undermined by AIDS', *South*, (1991), No. 123, pp. 17–18.

Pickering, H., 'Anthropological Methods Used in the Gambian MRC AIDS Research Programme', Paper presented to the conference: AIDS in Developing Countries: Appropriate Social Research Methods, (Brunel University, UK, 10–11 May 1990).

Piot, P. et al., 'Acquired Immunodeficiency Syndrome in a Heterosexual Population in Zaïre', *The Lancet*, (1984), Vol. 2, pp. 65–9.

Piot, P., Mann, J., 'AIDS in the tropics', *Baillière's Clinical Tropical Medicine and Communicable Diseases*, (1987), Vol. 2, pp. 209–22.

Preble, E., 'Impact of HIV/AIDS on African Children', *Social Science and Medicine*, (1990), Vol. 31, No. 6, pp. 671–80.

Public Eye, 'Zimbabwe Admits AIDS Increase', (17 May 1991).

Rajab, A., 'Tanzania: where the "insect" could destroy the tourist industry', *AIDS Analysis Africa*, (May/June 1991), Vol. 1, No. 2, p. 9.

Ranger, T., Slack, P., 'Epidemics and Ideas: Essays on the Historical Perception of Pestilence', *Past and Present Publications*, (Cambridge University Press, Cambridge, 1991).

Raynham-Small, M. R., Cliffe, A. D., 'Acquired Immunodeficiency Syndrome (AIDS): The Global Spread of HIV Type 2', in R. W. Thomas (ed.), *Spatial Epidemiology*, (1990), pp. 139–82.

Rees, M., 'Miners, refugees, soldiers at risk', *WorldAIDS*, (July 1989), No. 4, p. 6.

Richards, A. I., *The Changing Structure of a Ganda Village: Kisozi 1892–1952*, (East African Institute of Social Research, Kampala, 1966).

Richards, P. W., *The Tropical Rainforest*, (Cambridge University Press, Cambridge, 1952).

Rider, E., 'Let us fight it together', *WorldAIDS*, (September 1989), No. 5, p. 5.

Rodrigues, Lair Guerra de Macedo, 'Public Health Organisation in Brazil', in A. R. Fleming et al. (eds), *The Global Impact of AIDS*, (Alan R. Liss, New York, 1988), pp. 229–32.

Roscoe, J., *The Baganda*, (first published 1911, republished 1965, Frank Cass, London).

Rothschild, D., Harbeson, J. W., 'Rehabilitation in Uganda', *Current History*, (March 1981), Vol. 80, No. 463.

Rowley, J. T., Anderson, R. M., Wan Ng, T., 'Reducing the Spread of HIV infection in sub-Saharan Africa: some demographic and economic implications', *AIDS*, (January 1990), Vol. 4, pp. 47–56.

Sathyamurthy, T. V., *The Political Development of Uganda 1900–1986*, (Gower Publishing, Aldershot, 1986).

Schofield, R, 'Family Structure, Demographic Behaviour and Economic Growth', in J. Walter, R. Schofield (eds), *Famine, Disease and the Social Order in Early Modern Europe*, (Cambridge University Press, Cambridge, 1989), pp. 279–304.

Selik, R. M., Noble, G. R., Meade Morgan, W., Dondero, T. J., Curran, J. W., 'Epidemiology of AIDS and HIV Infection in the United States', in A. R. Fleming et al. (eds), *The Global Impact of AIDS*, (Alan R. Liss, New York, 1988), pp. 35–41.

Sen, A., *Poverty and Famines: An Essay on Entitlement and Deprivation*, (Oxford University Press, Oxford, 1981).

Serawadda, D., Katongole-Mbidde, E., 'Viewpoint: AIDS in Africa, problems for research and researchers', *The Lancet*, (7 April 1990), No. 335, pp. 842–3.

Serawadda, D., Musgrave, S., Wawer, M., Musagara, M., Konde-Lule, J.,

Sewankambo, N., 'Rakai Project–ACP', Unpublished mimeo, AIDS Control Programme, Entebbe, (1990?).
Serote, R., 'Trying to reach the men of Soweto', *WorldAIDS*, (May 1990), No. 9, pp. 5–6.
Sheon, A. R. et al., 'A National Survey of AIDS Awareness in Zimbabwe and Botswana and the Uses of Demographic and Health Survey Data for AIDS Program Planning', IVth Conference on AIDS and Associated Cancers, (Marseille, 1989).
Shilts, R., *And the Band Played On: Politics, People and the AIDS Epidemic*, (St Martin's Press, New York, 1987).
Skitovsky, A. A., Over, M., 'AIDS: costs of care in the developed and the developing world', *AIDS*, (1988, Supplement 1), No. 2, pp. S71–S81.
Small, N., 'AIDS and Social Policy', *Critical Social Policy*, (Spring 1988), Vol. 1, No. 21.
Southall, A., 'Social Disorganisation in Uganda: before, during and after Amin', *Journal of Modern African Studies*, (1980), Vol. 18, pp. 627–56.
Southall, A., 'The Recent Political Economy of Uganda', in H. B. Hansen, M. Twaddle (eds), *Uganda Now: Between Decay and Development*, (James Currey, London, 1988), pp. 54–69.
Southern African Economist, The, 'Frightening Costs of AIDS', (April/May 1989), pp. 53–4.
Specter, M., 'The Case of Dr Gallo', *New York Review of Books*, (15 August 1991), Vol XXXVIII, No. 14, pp.47–50.
Standing, H., Kisekka, M. N., *Sexual Behaviour in Sub–Saharan Africa: A review and annotated bibliography*, (Overseas Development Administration, Glasgow, 1989).
Stock, R., 'Disease and Development or the Underdevelopment of Health: A Critical Review of Geographical Perspectives on African Health Problems', *Social Science and Medicine*, (1986), Vol. 23, No. 7, pp. 689–700.
Sunday Times, The, 'Straight, Anglo–Saxon, and still at Risk', (London, 25 March 1990).
Sunday Times, The, (London, 1 June 1990).
Tadria, H. M. K., 'Changing Economic and Gender Patterns among the peasants of Ndejje and Ssegulium in Uganda', unpublished PhD Thesis, (University of Minnesota, 1985), Dissertation Abstracts 1986 46/07 p. 1993–A, pp. 63, 73, 92.
Tebere, R., 'Uganda Opens New Front', *WorldAIDS*, (March 1991), No. 14, p. 3.
Thomas, R. W. (ed.), 'Spatial Epidemiology', *London Papers in Regional Sciences*, (1990), No. 21.
Thompson, G., 'Lessons from an anthropological survey of traditional health practitioners and their role in the HIV transmission and its prevention', unpublished paper presented to the conference on AIDS in Developing Countries: Appropriate Social Research Methods, (Brunel University, UK, 10–11 May 1990).
Thorner, D., Kerblay, B., Smith, R. E. F., *Chayanov on the Theory of Peasant Economy*, (Richard D. Irwin, Homewood, Illinois, 1966).
Traore, A., 'The Cost of AIDS: food for thought', *The Courier*, (March–April, 1991a), No. 126, pp. 47–9.
Traore, A., Interview with Dr Michael Merson: 'We have to think of AIDS as a development problem, not just a health one', *The Courier*, (March–April, 1991b), No.126, p. 50–52.
Traore, A., 'AIDS The Big Threat', *The Courier*, (March–April, 1991c), No. 126, p. 42–6.

Tuju, R. and the Panos Institute, 'Kenya's Pied Crow gives the facts', *WorldAIDS*, (January 1990), No. 7, p. 4.

Tuju, R., 'A Brief Report on letter response to the radio programmes "AIDS the fact"', *Kenyan AIDS Newsletter*, (February 1990), Vol. 2, No. 2, p. 5.

United States Department of Agriculture, Economic and Statistics and Cooperative Services, International Economics Division, Africa and Middle East Branch, *Food Problems and Prospects in Sub-Saharan Africa - The Decade of the 1980s*, (September 1990).

Van de Perre, Ph., Rouvrey, D., Lepage, Ph., Bogaerts, J., Kestelyn, P., Kayihigi, J., et al., 'Immunodeficiency syndrome in Rwanda', *The Lancet*, (1984), Vol.2, pp. 62–5.

Van Zwanenberg, R. M. A. with King, A., *An Economic History of Kenya and Uganda, 1800–1970*, (Macmillan, London, 1975).

Vaughan, M., 'Syphilis in Colonial East Africa: the social construction of an epidemic', in T. Ranger and P. Slack (eds), *Epidemics and Ideas: Essays on the Historical Perception of Pestilence, Past and Present Publications*, (Cambridge University Press, Cambridge, 1991).

Walter, J., Schofield, R. (eds), *Famine, Disease and the Social Order in Early Modern Europe*, (Cambridge University Press, Cambridge, 1989).

Watson, C., 'Ugandan schoolgirls resist sweet talk', *WorldAIDS*, (November 1989), No. 6, p. x.

Watts, M., 'Coping with the market: uncertainty and food security among Hausa peasants', in I. de Garine, G. A. Harrison (eds), *Coping with uncertainty in food supply*, (Clarendon Press, Oxford, 1988).

Wellcome Foundation, *AIDS: A Global Health Crisis*, (The Wellcome Foundation, Berkhamsted, England, 1989).

West, H. M., *Land Policy in Buganda*, (Cambridge University Press, Cambridge, 1972).

Whiteside, A., van Niftrik, J., 'AIDS in South Africa: Government and ANC Response', *AIDS Analysis Africa, Southern Africa Edition*, (August–September, 1990), Vol. 1, No. 2, pp. 1–2.

WHO (World Health Organization), 'Epidemiologically based HIV/AIDS projection model', mimeo, (WHO, Geneva, 1988).

Wiebe, P. D., Dodge, C. P., *Beyond Crisis: Development Issues in Uganda*, (Makerere Institute of Social Research and the African Studies Association, Kampala, 1987).

Williams, G., *From Fear to Hope: AIDS Care and Prevention at Chikankata*, (Actionaid, Hamlyn House, Archway, London in association with AMREF and World in Need, 1990).

World Health Organization, 'Projections of HIV/AIDS in Sub-Saharan Africa using an epidemiologically based modelling approach', mimeo, (Geneva, 26 April 1989).

World Bank, *Financing Health Services in Developing Countries: An Agenda For Reform*, (World Bank, Washington, 1987).

World Health Organization, *Weekly Epidemiological Record*, Geneva, (4 May 1990), No.18.

*World*AIDS, No. 1, (January 1989), p. 10.

WorldAIDS, 'Larger than life communication', No. 4, (July 1989), p. 5.

WorldAIDS, No. 12, (November 1990), pp. 5–9.

WorldAIDS, No. 4, (July 1989), p. 5.

WorldAIDS, No. 14, (March 1991), p. 3.

WorldAIDS, 'Meeting reports condom acceptance', No. 13, (January 1991), p. 3.

WorldAIDS, No. 13, (January 1991), p. 6.
WorldAIDS, No. 13, (January 1991), p. 3.

Index